The Reality of American Energy

The Reality of American Energy

The Hidden Costs of Electricity Policy

Ryan M. Yonk, Jordan Lofthouse, and Megan Hansen

PRAEGER™

An Imprint of ABC-CLIO, LLC
Santa Barbara, California • Denver, Colorado

Library of Congress Cataloging-in-Publication Data

Names: Yonk, Ryan M., author. | Lofthouse, Jordan, author. | Hansen, Megan, author.
Title: The reality of American energy : the hidden costs of electricity policy / Ryan M. Yonk, Jordan Lofthouse, and Megan Hansen.
Description: Santa Barbara, California : Praeger, [2017] | Includes bibliographical references and index.
Identifiers: LCCN 2017007696 (print) | LCCN 2017024786 (ebook) | ISBN 9781440853920 (ebook) | ISBN 9781440853913 (acid-free paper)
Subjects: LCSH: Electric power production—Government policy—United States.
Classification: LCC TK1191 (ebook) | LCC TK1191 .Y55 2017 (print) | DDC 333.793/20973—dc23
LC record available at https://lccn.loc.gov/2017007696

ISBN: 978-1-4408-5391-3
EISBN: 978-1-4408-5392-0

21 20 19 18 17 1 2 3 4 5

This book is also available as an eBook.

Praeger
An Imprint of ABC-CLIO, LLC

ABC-CLIO, LLC
130 Cremona Drive, P.O. Box 1911
Santa Barbara, California 93116-1911
www.abc-clio.com

This book is printed on acid-free paper ∞

Manufactured in the United States of America

Contents

Acknowledgments

The authors would first like to thank everyone at Strata Policy and the Institute of Political Economy at Utah State University who helped contribute to this book, especially to the dozens of student research associates who spent countless hours researching and collaborating.

We further want to thank our significant others who put up with the long hours, endless discussions of electrical generation and energy policy. Without their support and patience this book would not have been possible. So to Bradley, Troy, and Cody we say thank you.

As we worked to put this book together, we were able to explore new and interesting ideas that have the potential to have huge implications for the environment, the electric grid, and as a result human flourishing. We hope that the people who helped contribute to this book and all those who read it will work to achieve more prosperous and free societies by affecting a change in the climate of ideas.

Introduction

Politicizing Electricity

Nearly every aspect of American life is politicized, even those we don't think much about. For better or worse, nearly everything we use is subject to the political push-and-pull of laws and regulations. From the sugar we put in our morning coffee and the corn we grow in our fields, to the electricity we use to turn on our lights, government has a hand in almost everything and the process by which government makes decisions is a political one.[1] Despite considerable consternation among the general public about politics, most people would agree that some government oversight over industry is necessary to protect human health and welfare. The reality of policy making, however, means that the public interest is not always served by the government officials who are sworn to protect it. Despite the good intentions of policy makers, they can skew markets and cause harm to consumers and citizens.

One of the clearest examples of the impact political decision making has on American consumers can be found in the inflated prices paid for sugar; federal policies artificially raise the price of sugar by imposing tariffs that make foreign sugar more expensive than domestically produced sugar and by providing subsidies to domestic sugar production. Despite claims by supporters of these sorts of policies that tariffs on imported sugar protect American jobs and keep economic activity at home, the reality is that a freer trade in sugar would benefit consumers in substantial ways.[2] As a result of the tariff policy, American sugar consumers are stuck paying the bill, both for higher taxes (to pay for the subsidy) and higher sugar prices (because the market is restricted).[3]

The sugar industry is one of the most commonly cited examples of the results of political decision making that picks winners and losers in the marketplace. Impacts of this sort are not, however, limited to the sugar market. They extend across large sections of the energy sector. The gasoline market, for example, has seen federal policy makers persuaded by corn ethanol producers to provide billions of dollars in subsidies and mandate ethanol use in gasoline. Proponents of ethanol subsidies and mandates argue that we should produce fuel from corn because it is a renewable resource and, presumably, its environmental impact is smaller than ethanol-free gasoline. Ethanol producers are awarded money from taxpayers who are forced to pay more through both higher gasoline prices and taxes that fund massive subsidies to corn farmers. Ethanol producers win, but the American public loses.[4]

Just like sugar producers, corn producers, or any other company, electricity producers often try to influence the political system for their own benefit. Businesses know that it is profitable to have the government on their side. Many business leaders try to convince government officials to pass laws and regulations on their behalf, which can limit competition and grant privileges and additional revenue to the lucky corporations who gain the favor of policy makers. When the government grants special power to certain corporations or industries, the favored few are better off and their competitors and ultimately consumers are worse off.

Beginning in the early 20th century, fossil fuel industries persuaded policy makers to enact special benefits for companies who supply coal and natural gas. The federal government and many state governments have extended subsidies, tax breaks, and other privileges to the fossil fuel corporations who were the most persuasive and who were best able to use the political process to their advantage. More recently, that support has begun to dwindle. Interestingly, while subsidizing fossil fuels, policy makers simultaneously make them more expensive in response to political pressures from other special interest groups (namely environmental groups).

Climate change and environmental degradation have become highly politicized topics, and with this political fervor, renewable energy corporations have found strategies similar to those used in the fossil fuel industry to be effective ways to capture politically provided privileges, commonly call rents. Advocates for renewable energy sources have successfully lobbied the government for special treatment. They argue that more renewable energy will be good for the environment and create thousands of jobs. Renewable energy companies now enjoy tens of billions of dollars in subsidies every year. On the state level, special interest groups for renewables have been able to make deals with policy makers to their own benefit. Now renewable energy sources are mandated by law in more than half the states, regardless of the cost or reliability of these energy sources.

Mixing politics into the energy sector, or any sector, is a messy business. Government officials have the power to pick winners and losers, so competing interest groups work to secure the best outcome for themselves while limiting their competition. The motivations for, and the outcomes from, government regulations are complex. This book is meant to show who makes electricity policies, how they make them, and what the actual outcomes are. At first glance, many electricity policies may seem prudent or even noble, but the reality of energy policy is not as clear-cut or beneficial as at first glance it may appear.

The Reality of Electricity Policy

The purpose of this book is to provide a new way to look at the realities of electricity policy in the United States. We use public choice, a branch of political economy, to analyze the major electricity sources in the country. Public choice theory asserts that individuals are self-interested actors who make decisions in an effort to make themselves better off. Government officials and leaders are not exempt from this principle, but their self-interested decisions often have far-reaching impacts on the American public. Public choice theory can help us explain a great deal of why good intentions often lead to poor policy outcomes, especially in the area of electricity policy.

Each electricity source has unique pros and cons that are easy to overlook. For example, solar power is more effective and efficient in the sunny Southwest than in snowy Minnesota. Hydropower makes more sense in the Pacific Northwest where there is an abundance of large rivers and consistent rainfall. Informed decisions about electricity production require accurate information about how each electricity source works and the costs and benefits associated with each.

When government policies alter electricity markets, electricity producers and consumers change the way that they make decisions regarding which electricity sources to use and how much to produce. Producers and consumers begin to weigh the pros and cons of each energy source differently. Without invasive government policies, producers and consumers make electricity-related decisions based on the sources that are most economical and fit their preferences. Regardless of the intentions of the policy makers, laws and regulations make energy sources cheaper or more expensive, distorting how producers and consumers choose to respond. These policies also muffle the market signals about what people prefer, meaning that the main signal that producers respond to is what policy makers want, rather than consumers. Public choice can shed light on the motivations behind energy policies and explain how and why actual outcomes often differ so widely from intended outcomes.

The 12 chapters in this book explain how electricity is generated and transmitted, how public choice economics can explain electricity policy in general, and how public policies affect the most common electricity sources.

In Chapter 1, we discuss how electricity is generated, transmitted, and consumed. Understanding the basics of how electricity gets from power plants to homes and businesses is essential to understanding how government policies affect energy industries today. Policy makers and the general public can make more informed decisions about our country's energy future by first gaining an understanding of how the electricity system works physically and politically.

The physical process of generating, transmitting, and consuming energy is important, but the economics, politics, and environmental impacts of electricity are just as important as the physics of electricity. In Chapter 2, we explore the political side of how laws and regulations are made and the incentives of the policy makers. In Chapter 3, we examine the economic implications of how those laws change the way people behave and distort energy markets. Politicians, bureaucrats, and special interest groups use the power of government to get what they want, so understanding how the political process functions is a necessary foundation for understanding how energy policy is formed. Economics is the study of how people make decisions and exchange with one another, which is key to understanding how government policies shift the way that people make decisions and exchange with one another.

In Chapters 4 through 11, we explore each of the major electricity sources in the United States: wind, solar, biomass, geothermal, hydropower, nuclear, coal, and natural gas. Each electricity source has unique pros and cons. For example, coal-powered plants produce large amounts of reliable electricity, but they also emit harmful pollutants. Solar and wind power do not produce air pollution directly, but because the sun does not always shine and the wind does not always blow, they have serious reliability issues that cannot be ignored. Like coal power plants, hydropower dams can generate large amounts of consistent electricity, but constructing dams alters the ecology of whole river systems. We explore how each energy source works and how government policies have altered the markets for each. Government policies may try to increase the benefits of certain energy sources, but these policies can have hidden costs and unintended consequences. We highlight how government policies have helped or harmed the electricity production industry, and how the unintended consequences of these policies have imposed hidden costs on the American public. In Chapter 12, we conclude this book with an exploration of the realities of American energy and the policies that shape it.

CHAPTER ONE

Understanding Electricity

Basics of Electricity Production

Electricity is fundamental to everyday life in the United States. Most people, however, do not know where their electricity comes from. We flip the switch and expect the lights to turn on. Making sure that the lights always turn on is far more complicated than most people realize. Interacting systems of electricity generation, transmission, and consumption allow efficient distribution of electricity that has allowed us to live comfortably and with little concern about whether power will be there when we flip the switch and desire it.

Electricity is simply moving electrons that we can harness to do work. Power plants and electricity generators get electrons moving, and then we transport those electrons through "the grid." The grid is the interconnected web of electricity generators, transmission lines, converters, and electricity consumers. Everything required to move electricity to those who need it is part of the grid.

Electricity generators come in many forms. Fossil fuel power plants produce the most electricity in the United States. As of 2015, both coal and natural gas each generated 33 percent of U.S. electricity, for a combined total of about two-thirds of the electricity we consume. Nuclear power generates about 20 percent of our electricity, and all renewable energy sources combined produce about 13 percent.[1]

Transmission lines move electricity from power plants to the homes and businesses that use it. When electricity first leaves a power plant, it is sent through a transformer, which increases the voltage, making it more efficient for long-distance transmission. As the electricity nears its destination, it goes through more transformers that lower the voltage to a usable level for households.

The grid is actually a complex system of smaller, loosely interconnected grids. Electricity is transmitted through this network of smaller grids that are each owned by different providers. Publicly owned utilities, investor-owned utilities, cooperatives, federal power agencies, and power marketers all provide American homes and businesses with electricity. Investor-owned utilities serve the most customers in the United States with 68 percent of the market. These utility companies have their own power plants to generate electricity and have infrastructure to send electricity to consumers. Sometimes power plants are independently owned rather than owned by a utility company that transports electricity to homes and businesses. Private generating companies not owned by a utility company produce nearly 40 percent of the nation's electricity.[2]

An independent power producer can sell electricity to the utility company or other electricity buyer. A power purchase agreement (PPA) is a contract between independent power producers and utility companies. Electricity sellers agree to provide certain amounts of electricity to a utility company, usually for 5 to 20 years, at an agreed upon price. Because each circumstance is unique, each PPA is unique depending on the type of energy source, the needs of the utility company, and electricity demand in a certain area. PPAs help reduce uncertainty for power producers and utility companies because electricity producers known how much revenue they will make, and utility providers do not have to worry about the upkeep of power plants.[3]

The utilities, cooperatives, and agencies that generate and provide power don't work alone. The Federal Energy Regulatory Commission (FERC) regulates and oversees electricity providers. FERC has also set up entities to help facilitate electricity transmission over larger areas, including regional transmission organizations (RTO) and independent system operators (ISO) that coordinate and control how electricity is transmitted. RTOs and ISOs function in similar ways, but RTOs generally cover a larger area. They are independent of power markets and oversee grid operations, short-term reliability, and transmission service.[4]

Power plants, utility companies, system operators, and government agencies all work together to make sure that electricity is consistently and reliably sent to consumers. Whenever a light switch is flipped on, hundreds of people working at different jobs in the grid system make sure that electricity is available.

One of the most important facts about the grid is that electricity supplied must equal electricity demanded at any given moment. At the grid level, there is no way to store electricity, so any electricity that is fed into the grid must be consumed simultaneously. Grid managers and their computers work constantly to maintain the balance between the incoming electricity from power plants and the electricity consumption from homes, schools, and businesses. When there is not enough electricity, grid managers need to tell power plants in real time to produce more energy to maintain the balance. When there is too much electricity, grid managers have to curtail energy generation at some power plants to make sure that there are no power surges. Grid managers cannot send electricity to a specific place. They can only make sure that there is exactly enough electricity entering local grids as people taking out.

Brownouts and blackouts occur when electricity supply and demand don't match up. Blackouts happen when grid managers have to shut off sections of the grid to keep the supply of electricity stable, or when a disruption to the grid cuts off power. Brownouts happen when there is not enough electricity in the grid. Grid managers may also intentionally cause brownouts to avoid complete blackouts. Fortunately, despite their relative commonality in the rest of the world, blackouts and brownouts are rare in the United States because grid managers are generally capable of keeping the electricity supply stable.[5]

Each Electricity Source Works Differently

Not all electricity sources work the same way. For some energy sources, producers can control how much electricity they produce, but for others, they cannot. For example, electricity producers can choose to burn more or less coal to meet electricity demand. Coal is "dispatchable" because electricity producers can choose to increase or decrease the output of a coal power plant to meet demand. Most energy sources are dispatchable, including natural gas, nuclear, hydropower, geothermal, and biomass. Although the exact amount of output at dispatchable energy sources can be controlled, some kinds of power plants take more time to adjust their output. For example, some types of natural gas power plants can change how much electricity they produce in a few seconds to a few minutes. Other types of power plants, like coal and nuclear, may take about 15 to 30 minutes to reach the desired level of output.

Most kinds of wind and solar power, however, are nondispatchable because they are dependent on favorable weather conditions. If the wind stops blowing, electricity producers cannot make the wind blow. If clouds block the sun or the sun sets, solar plants can no longer generate electricity. Nondispatchable power sources make it more complicated for grid managers to ensure that electricity supply and demand remain in balance. In areas with large amounts of nondispatchable power sources, grid managers cannot precisely control electricity supply to demand, which means that they have to be ready to call on dispatchable sources to ramp up output if there is a shortage of electricity or ramp down output if there is a surplus.

Some electricity sources are better at providing large amounts of consistent, reliable electricity than others. Power plants that provide large amounts of electricity are known as baseload power sources because they generate a "base" of electricity that we can use throughout the day, regardless of smaller fluctuations in demand. Baseload power sources work best when they maintain a relatively constant output. They are not meant to make wide swings in electricity output quickly. Coal and nuclear power plants are some of the most common baseload power plants in the nation. These electricity sources consistently produce large amounts of electricity to fit large-scale electricity consumption patterns throughout the day. For example, electricity demand will almost always be higher at 7:00 p.m. than at 3:00 a.m., so baseload power plants raise output to meet the majority of electricity demand at 7:00 and decrease output in the middle of the night. Baseload power plants are almost always dispatchable power sources because grid managers can control output to follow the general trends of electricity demand as they change throughout the day.

Although baseload power plants meet large-scale patterns of electricity demand, the exact amount of electricity that is needed changes minute by minute. To meet these smaller fluctuations, other energy sources serve as peak power sources, which change their output quickly to meet demand exactly. In most places, energy demand spikes in the morning and then again in the evening. These spikes in demand can happen in just a few minutes, and they may only last for a few minutes. Baseload power plants, like coal and nuclear power plants, are dispatchable, but they are not designed to change their outputs in short amounts of time. Other dispatchable sources, like natural gas and hydropower plants, are usually the best options for meeting these peaks in electricity demand quickly and efficiently. Natural gas can be burned in combustion turbines that work much like jet engines, which can generate electricity almost instantly. Hydropower also works well as a peak power source because additional water can easily be released through the hydroelectric turbines in a dam, producing more electricity almost instantly. Hydropower, however, relies on there being enough water ready when electricity is demanded. In years of drought or low water levels, hydropower capacity may be reduced. Wind and solar, because they are nondispatchable, cannot reliably serve as a baseload power sources, and are poor peak power sources.

Every power plant is a different size with a different output. People in the electricity industry often talk about the "capacity" of a power plant, which is usually measured in megawatts. In the most basic terms, a watt is measurement for an amount of electricity at any moment. A kilowatt is a thousand watts, and a megawatt is a million watts. Most people have a difficult time visualizing how much a megawatt is. To the average person, it is difficult to tell if a power plant with a capacity of 500 megawatts is big or small. A typical coal-fired power plant, for example, will have a capacity of 500 megawatts. A single onshore wind turbine may have a capacity of 1 to 3 megawatts, depending on its size. If a coal power plant has a capacity of 500 megawatts, then the most electricity it could produce at any given moment is 500 megawatts, assuming that it is producing at its full potential.

Capacity alone does not communicate how much electricity a power plant actually produces. Megawatt hours are used to measure the amount of electricity a power plant produces over time. For example, if a 500-megawatt power plant produces electricity at its full potential for one hour, then it will generate 500 megawatt-hours. If, for some reason, the same power plant is producing at only half its capacity for one hour, then it will generate 250 megawatt-hours. The number of megawatt-hours that a power plant produces depends on how much electricity is demanded and how well the power plant can meet its full potential. In 2015, all the commercial power plants in the United States generated about 4 billion megawatt-hours of electricity.[6]

On a monthly electricity bill, each household or business pays for the number of kilowatt-hours they consume. In 2014, the average American home used 10.9 megawatt-hours for the year. The monthly average for an American home is just under 1 megawatt-hour. People in different places across the country, however, use different amounts of electricity. Louisiana, for example, is the most electricity-hungry state, where the average household consumed 15.5 megawatt-hours in 2014. Hawaiians used the least amount of electricity, consuming only 6 megawatt-hours per household.[7]

The capacity of a power plant can be misleading because each energy source generates electricity differently. Even though a 500-megawatt coal power plant and a 500-megawatt wind farm have the same capacity, they will not produce the same amount of electricity. Coal power plants are used as baseload power sources because they consistently produce large amounts of electricity, whereas wind power is variable and intermittent because the wind does not always blow. Even with the same nominal capacity, the actual number of megawatt-hours that a wind farm generates will be much smaller than a coal power plant because wind cannot be controlled.

The term "capacity factor" is used to describe the potential output of a power plant compared to the actual output. If a power plant had a capacity factor of 100 percent, that power plant would be operating at its full capacity every second of every day. If a power plant had a zero percent capacity factor, that would mean a power plant was never working. Some power plants have a much higher capacity factor than others. Onshore wind farms, for example, have a capacity factor ranging from 30 to 40 percent, depending on the location and which types of turbines are used. Photovoltaic solar panels range from 15 to 30 percent, on average. On the other end of the spectrum, coal power has a capacity factor of 75 to 95 percent, and nuclear power ranges from 85 to 90 percent.[8] A clear understanding of these production realities is core to effective energy policy. Policy decisions to rely on sources with comparatively low capacity factors require over investment to ensure that the necessary electricity is available in real time.

Lifetime Costs and Environmental Impacts

It is difficult to accurately quantify how environmentally friendly energy sources are and how cost-effective they are. Two different types of analyses try to answer these two questions. First, life cycle assessments (LCAs) are meant to show how environmentally friendly each energy source is. Second, levelized cost of electricity (LCOE) estimates are meant to show how cost-effective each energy source is. Despite the purpose of these two types of analysis, they have significant shortcomings, and when policy makers and the public do not realize those shortcomings, they can make misinformed decisions.

Because environmental preservation is a driving force for many energy policies, government officials often use LCAs to evaluate the lifetime impacts that each electricity source has on the environment. These analyses begin with manufacturing and end with disassembly and recycling.[9] Ideally, LCAs should show us if the environmental impacts of each energy source are worth the costs. Organizations across the world have performed them on nearly every energy source to pinpoint the net environmental effects. As with any tool, LCAs are not perfect, and in many cases, they do not convey the full picture of how each energy source performs economically or environmentally.

Although LCAs may seem like a good way to critically analyze how well each electricity source can help or hurt the environment, they do not and cannot provide a comprehensive method of measuring how each energy source impacts the environment. First, individual LCA calculations employ different methodologies, meaning that conclusions about the same energy source can vary widely. Second, scientists often base these models on idealized assumptions about energy sources, not real-world data or actual performance. Third, LCAs simply cannot consider or quantify every single aspect involved in an energy source's life. No model can do this, however, and it is unreasonable to demand perfect, comprehensive accuracy. As a result, policy makers may make misinformed decisions if they assume that these analyses represent a fully accurate representation of the energy source.

As we noted, different LCA methodologies for the same energy source can yield wildly different results. For example, Anders Arvesen and Edgar G. Hertwich, scientists at the Norwegian University of Science and Technology, analyzed the methodologies that 34 different international institutions used to calculate LCAs for wind power. The only commonality among the 34 separate LCA methodologies was the identification of a manufacturing stage of a turbine's life span.[10] Even this commonality is not particularly meaningful as the emissions and energy impacts of manufacturing vary based on where and how turbines are manufactured. The manufacturing process is simply more environmentally costly in some locations compared to others.

Perhaps the most problematic part of life cycle assessments is that they are based on idealized assumptions rather than real-world data. As a result, they are less useful when trying to inform good public policy. For instance, researchers often use the name-plate life span of a wind turbine (20 years) in their calculations; however, real-world data show that many wind turbine life spans are closer to 12 to 15 years.[11,12] Recalculating LCAs with a 12- to 15-year life span would significantly lower the estimates of how much the wind turbines would actually reduce carbon emissions. As this example illustrates, real-world data are more useful when making calculations because the data provide a more accurate way for policy makers and the public to weigh wind power's overall environmental impacts. Using idealized models can easily mislead people into thinking that an energy source is better for the environment than it really is.

Even if LCAs have robust methodologies and use real-world data, they still face substantial problems. They cannot factor in every possible environmental cost related to the life cycle of an energy source. The entire process of manufacturing, transportation, decommission, and recycling makes it impossible for one analysis to incorporate all environmental impacts. Because LCA analyses do not include the environmental impacts of these processes, they likely underestimate the true environmental impacts of various energy technologies. LCAs may be a beneficial tool to help understand how the variation in different energy sources affect the environment, but they have serious shortcomings that policy makers should not disregard.

The levelized cost of electricity is another tool that policy makers and analysts use to estimate the average cost of producing electricity over a plant's lifetime. For example, the Transparent Cost Database that calculates the cost of producing electricity from solar panels is $290 per megawatt-hour.[13] The Energy Information Administration, on the other hand, found that the LCOE for solar panels is $125.3 per megawatt-hour.[14] The disparity between calculations for the same energy source is due to the fact that each organization calculates an LCOE by incorporating different data, which can skew the estimates significantly. LCOE estimates can't account for every cost in producing electricity, and some costs are difficult to see or measure. These costs may include reduced reliability of electricity production, tax dollars for government subsidies, or regulations. LCOE estimates can be useful as a tool to guide decisions, but policy makers and the public should realize that these estimates do not contain all the information that many assume they do. Understanding the actual cost of producing electricity from each energy source requires considering the many hidden costs involved in electricity generation.

Crafting good public policy is a difficult business because identifying what is actually in the best interest of the public is not a clear-cut matter. Politicians and bureaucrats often get to decide what is in the public interest, and in recent decades, the public benefit has come to primarily mean environmental quality. Government officials at the federal, state, and local levels have tried to encourage certain types of electricity sources that they view as environmentally friendly while discouraging the sources that they view as harmful. For many politicians and bureaucrats, any renewable energy source is automatically considered environmentally beneficial and deserving of governmental support. Other officials identify a different definition of "good" and try to promote the economic interests of fossil fuel industries that have existed in many states for decades. They say that they have the duty to protect the workers in their states, and by hurting the fossil fuel industry, they would also cause harm to the hard-working Americans they serve.

In politics, the line between altruism and selfishness is blurred, making it difficult to distinguish the actual motivations of policy makers. Politicians and bureaucrats may have genuine concerns about the environment and the economy, but they can use the same rhetoric to benefit themselves. Regardless of how well-intentioned or ill-intentioned public policies are, the actual outcomes of the policies are what affect real lives of real people. In the remaining chapters we explore what motivates policy makers to make particular policy decisions, how those decisions distort electricity markets, and the actual outcomes of U.S. electricity policy.

CHAPTER TWO

Regulating the Electricity Sector

Unlikely Allies: Bootleggers and Baptists

Until 2011, Georgians were not allowed to legally buy alcohol on Sunday anywhere in the state. Several years of debate led to a referendum in 2011 where each city in the state could decide whether to remain dry or to allow Sunday alcohol sales. Of the 127 cities that voted on whether to end the ban, 105 succeeded in ending it. Some of the most rural and religious areas of the state chose to keep the Sabbath day alcohol-free.[1]

The moral call for banning alcohol goes back to the temperance movement before the Civil War, and it still continues today.[2] In many places throughout the South, Baptists would lobby local and state government to pass ordinances restricting the ability of stores to sell alcohol. By eliminating liquor sales, they thought their communities would become more pious, making everyone better off.

Even more recently, parts of South Carolina are still having political debates over alcohol sales on Sunday. Business in the unincorporated portions of Greenville County are not allowed to sell alcohol on Sundays, but six incorporated municipalities within the county have already voted to allow alcohol sales. The voters went to the polls in November 2016 and voted to allow Sunday sales.[3] David McCraw, a restaurant owner, said during the election, "The businesses here in the unincorporated part are unable to compete head to head with business in the cities that surround us." The Baptist Ministers Fellowship of Greenville and Vicinity opposes the referendum in Greenville County for moral reasons.[4] The tension over alcohol sales remains a hurdle for the heavily Baptist South, considering that the Southern Baptist Convention seeks to limit alcohol use via legal means. The convention has formally stated that they "urge Southern Baptists to take an active role in supporting legislation that is intended to curb alcohol use in our communities and nation."[5]

When Sunday alcohol bans were more prevalent, bootleggers agreed with the Baptists who wanted to limit alcohol sales, albeit for different reasons. Bootleggers realized that people who wanted to buy liquor on Sundays would get it somehow, regardless of whether some local ordinance prohibited it. If stores could not sell alcohol on Sundays, the bootleggers did not have competition for at least one day a week. Without competition, the bootleggers could make more money from those people who would not give up the bottle on the Sabbath.

Economist Bruce Yandle was the executive director of the Federal Trade Commission (FTC) in the early 1980s. While working at the FTC, he noticed that many unrelated special interest groups became strange bedfellows when they agreed on particular issues. In other words, special interest groups intentionally or unintentionally form partnerships to achieve the same government policies. These groups, however, often benefit from the same policies in very different ways. Yandle, a southern gentleman himself, recognized that the push to ban alcohol on Sundays by southern bootleggers and Baptists was similar to the cooperation between unrelated special interest groups that he had seen while working in the FTC. His influential "bootleggers and Baptists" theory explained how special interest groups can and do work to get a certain law passed, even if these groups have different end goals.[6]

Yandle asserted that the bootleggers benefit because Baptists do the bulk of the campaigning and the persuading, but the bootleggers get to reap the rewards of their hard work. Baptists use religious or moral convictions to argue for their preferred law. The bootleggers then profit because the Baptists enforce a cartel-like agreement on Sunday alcohol sales. Bootleggers generally have less-than-noble motives and advocate for their preferred policies quietly, or they try to mask their true intentions by mimicking the moral arguments of the Baptists. Baptist-bootlegger coalitions in any industry make regulations more long-lasting, as long as the metaphorical Baptists and bootleggers have a common objective.

Many industries have a Baptist-bootlegger relationship, and some of these relationships are a bit stranger than others. The recent successful initiative to legalize marijuana in California is a clear example of how bootleggers and Baptists work together and what happens after their coalition falls apart. People who oppose marijuana because of alleged harms to health and safety could be considered the "Baptists" in this situation. Marijuana farmers in northern California, particularly Humboldt County, are one example of bootleggers because they want to keep marijuana illegal for their own financial benefit. When they grow and sell marijuana outside the law, they make large profits without paying taxes.

Ordinary business owners in northern California are another faction of bootleggers because the underground marijuana industry stimulates the local economy. Legalized marijuana would cause the illegal marijuana growers to lose their market share, and the local businesses would be harmed because one of their major draws to their community would shrink. It seems strange that local businesses oppose legalization because marijuana would be taxed and potentially provide more revenue for local communities. At first glance, local businesses would benefit from more people coming to buy legal marijuana than illegal, but law enforcement officials in some towns in northern California rarely enforce anti-marijuana laws, and legalization would hurt those who would be taxed more than it would help.

Other bootleggers include the alcohol industry and big pharmaceutical companies, which have both donated to anti-legalization efforts. With medicinal marijuana, pharmaceutical companies lose some of their market share, and with recreational marijuana, alcohol companies lose some of theirs.

In recent years, the moral arguments for keeping marijuana illegal have become much less accepted by the public. Some places beyond California, including Alaska, Colorado, Oregon, Washington, and the District of Columbia as well as (likely) Nevada, Maine, and Massachusetts have legalized recreational marijuana as well. Prior to the 2016 election an additional 20 states had legalized medicinal marijuana, and that total appears to have increased following successful ballot initiatives in a number of other states.[7] As the marijuana example shows, when the bootleggers lose their Baptists, the regulations that protect them are often quickly phased out.

The relationship between the actual Baptists and bootleggers in Yandle's example or those benefiting from marijuana prohibition is found across many industries albeit with less clarity as to who the Baptists and the bootleggers are. The electricity sector is not exempt from Baptist-bootlegger relationships and their effects. For example, many environmentalists, playing the part of the Baptists, push for policies that are friendly to renewable energy sources because they see renewables as a responsible and ethical way to improve environmental quality. Other groups, including renewable energy corporations and their suppliers, play the part of the bootleggers. These groups can take advantage of the hard work and money that environmental groups invest into lobbying. Of course, renewable energy corporations also lobby for preferential treatment, but they do not face the full costs of lobbying because environmental groups conduct significant portions of the lobbying around renewable energy. By riding the coattails of environmentalists, these bootleggers profit from government-granted privileges, such as massive government subsidies.

Further, in environmental issues, bootleggers often masquerade as Baptists. The federal government offers billions of dollars in subsidies each year to renewable energy corporations. Wind and solar power producers would be foolish to pass on "free money," so they often use the rhetoric of environmentalism to reap economic benefits. Renewable energy companies claim a moral high ground to persuade policy makers for continuing to supply government money.[8]

Environmental "Baptists" are also joining forces with low-carbon energy corporations to restrict carbon emissions and institute a cap-and-trade system. They argue that carbon emissions will cause economic and environmental harm, and that this harm justifies government regulations. Low-carbon energy companies, however, can make more profit in a restricted energy market. The nuclear industry, natural gas industry, and renewable energy industries generally support cap-and-trade policies because they would have less competition from the coal industry, which has historically been the top U.S. electricity source. By limiting competition from their biggest competitor, these other electricity industries have a compelling financial incentive to lobby for a cap-and-trade system. Brokers are necessary for a mandatory government regime of buying and selling permits and licenses under a cap-and-trade system.[9] Brokerage firms that can make a profit by facilitating the cap-and-trade system serve as another bootlegger in this scenario.

The formation of Baptist and bootlegger coalitions and how the two disparate groups come together is widely debated in the larger literature.[10] Do the bootleggers just take advantage of the Baptists' naiveté? Do they actively cooperate? Must a charismatic figure in the form of a political entrepreneur show the parties the possibilities a coalition can have?[11] It is likely that all of these can lead to various coalitions, but regardless of the mechanism all these relationships are rooted in the realities of American politics.

Public Choice Takes the Romance Out of Politics

A strange contradiction exists in the minds of many Americans. If you asked an average person on the street to describe a politician in a single word, he might say slimy, selfish, or power-hungry. Popular televisions shows, like the drama *House of Cards* and the comedy *Veep*, highlight the stereotype that politicians are self-serving and untrustworthy. Public opinion polls seem to confirm the stereotypical perceptions that many people have. Between 2006 and 2016, Gallup has found that congressional approval rating has been between 9 and 39 percent.[12] Yet, for all the distrust and disillusion surrounding politicians, some people have a tendency to separate "the government" as a whole from the individual elected officials who compose it. It is strange to assume that the government is somehow more benevolent than the politicians who make it up.

James M. Buchanan, an early founder of public choice theory, described this approach as "politics without romance." In other words, public choice sets up a framework to analyze the reality of how and why government officials make decisions. One of public choice's contributions to the social sciences is the focus on methodological individualism, which is a way to explain social phenomena by examining the motivations and actions of individual people. Although social scientists use many ways to understand how governments works, public choice offers another perspective of analyzing what motivates government officials and how they choose to behave. Public choice helps us look at the specific individuals in government and why they choose to do what they do. Using public choice to analyze how the government actually functions removes the magic and mystery from the policy-making process. Much like when Dorothy from *The Wizard of Oz* learned the truth, public choice allows us to see the man behind the political curtain.

The government in various forms has been involved in the electricity sector for about as long as Americans have been using electricity. Politicians have helped decide what kinds of electricity we produce, who should produce it, and how much it should cost. These decisions are not scientific decisions made with unbiased data. Many decisions regarding electricity have been, and continue to be, politically motivated and strategically crafted to benefit people in power. Throughout this book, we show how politics and electricity have become interconnected, providing a logical and realistic way of interpreting the history of energy policy in the United States.

First, public choice rejects the idea that government officials work only for the public good. Economists consider humans as rationally self-interested utility maximizers. In the language of the noneconomist, people do what makes them happy. Pinning down a definition of happiness is tricky because everyone has different preferences, and those preferences are subjective. Economists assume that people rank different options in a way that is consistent with their preferences. After people rank their preferences, they make goals to achieve them. Despite the wide variety of goals and desires, people generally have similar overarching desires. People often want wealth and power, but the forms of wealth and power can be diverse. People also want to make the world be how they think it should be.

Politicians and bureaucrats are just like other people.[13] Public choice theory argues that government officials are not magically transformed into enlightened rulers who do what is best for society. Politicians and bureaucrats take actions that they believe will benefit them. Sometimes what benefits them benefits society as a whole, and sometimes it harms society. As outside observers, the public may see politicians as having nefarious or selfish motives, but the reality of how government officials behave is incredibly complex. Even if politicians only wanted to do what was good for those they govern, they would still need to seek reelection to accomplish those goals. If they are not concerned with reelection, they are selected out of the political system. One suggestion to overcome the problem of politicians being self-serving is simply electing "good" people, but reality is not nearly that simple. The system itself, through a process of natural selection, creates these selfish behaviors, which forces even "good" politicians to behave selfishly.

One of the key forces pushing this behavior is special interest groups. For conspiracy theorists, the mention of special interest groups might conjure images of conniving lobbyists making deals in smoky back rooms with corrupt politicians. A more realistic approach sees special interest groups as communities of people who work to promote a specific preference. Exxon-Mobil, Mothers Against Drunk Driving, WildEarth Guardians, and the National Coalition for the Homeless all lobby politicians for what they want. Whether a special interest group is good or bad depends on what an individual thinks is good or bad for the world. For instance, people often complain about other special interest groups, while simultaneously supporting groups that advance their own interests.

When special interest groups and politicians collude, taxpayers are left holding the bill. If an oil company lobbies for a subsidy and Congress passes a law granting the subsidy, the American taxpayers are forced to pay that oil company. Paying taxes is clearly not negotiable; violating tax law can lead to fines or imprisonment. Lying on tax forms or choosing to not pay taxes is clearly a gamble with serious consequences. Average citizens do not get to choose how much they spend in taxes or where their tax dollars go to. Politicians, on the other hand, may freely choose how to spend other people's money with minimal consequences or oversight.

Politicians, like all people, are complex, and distilling their motivations down to a single source is difficult if not impossible. Even though politicians have a wide set of preferences, there are a few common factors that motivate them. Most want to be reelected and maximize their power. To help them do this, they work with special interest groups because these groups have money and influence. Politicians need campaign funds and ultimately votes, and special interest groups efficiently provide both funds and often the votes needed to improve the likelihood of reelection. Politicians can provide the special interest groups a path to their preferred policy outcome. Thus working together makes sense for both the special interest groups and politicians.

This cooperation leads to special privileges for those groups and certain companies or industries. In essence, politicians have the power to choose winners and losers in any industry at the taxpayers' expense. Special interest groups devote time and money to being picked as the winner, and if they are, they donate funds and persuade voters to increase the chance that friendly politicians are reelected. If a politician were to deny a special interest group what they want, the group may actively try to persuade people to vote the current politician out and vote in someone who is more sympathetic to their cause.

When policies grant special benefits, those benefits are concentrated among the few companies who reap the rewards of the subsidy or tax break. These companies have an incentive to lobby to keep favorable policies in place. The costs, however, are dispersed among all taxpayers. For example, a company may receive $50 million a year in government subsidies, but the subsidy costs each taxpayer $1 each year. The majority of taxpayers may disapprove of the favoritism that is extended to that particular company, but the costs outweigh the benefits for trying to fight against such favoritism. In other words, it would cost an individual taxpayer much more than $1 to devote the time and resources to persuade policy makers to change the policy. Concentrated benefits and dispersed costs are one explanation why seemingly unfair or wasteful special privileges, subsidies, or other preferential treatment persists.

Passing Laws and Issuing Regulations

After politicians make a law, they pass responsibility to government agencies to implement the law. These agencies and the bureaucrats who work in them are overseen by the president, as the head of the executive branch. Ideally, the job of agencies is to make sure the laws that Congress passes do what they are meant to do. But just like politicians, bureaucrats are self-interested and respond to incentives. They try to use their positions to further their preference and worldviews. Although bureaucrats are complex with many motivations, a few general motivations are common. Bureaucrats make decisions that increase their budgets, power, and influence. In many cases, regulators have strong ideals regarding the things they regulate, which may blind them to other considerations. They often make decisions that affect the whole country based on single issues while not considering other concerns.[14] When regulators are passionate about the things they regulate, they may fail to consider anything they think is irrelevant to their mission, although these other things have real economic, social, and environmental implications.

Rules and regulations are one of the most common ways that agencies fulfill the requirements of a law, but the rulemaking process is not problem-free. In many cases, rulemaking begins without a concrete direction. Bureaucrats do not always gather ample evidence to justify a proposed rule. They do not always have a detailed reason why they want to regulate a particular thing in a particular way. Rules and regulations are the way that the executive branch fulfills the goals of legislation, but sometimes regulators do not explicitly state why their regulation is necessary or how it fulfills the specific provisions of a law. Regulators may lack an articulate theory as to why a regulation would be better than simply allowing the market to act without interference.

Even in the best cases when agencies use evidence to support their reasoning for a regulation, the evidence is often unreliable. Anecdotes, internal agency expertise, and expert testimony are the basis for many regulations. But one of the problems with basing nationwide regulations on these types of evidence is that they cannot be tested easily, if at all. If an agency issues a regulation without rigorous testing before it is imposed, it could have substantial impacts, and unintended consequences may be felt throughout the country. For example, a government agency, such as the Federal Trade Commission, has justified their proposed regulation by collecting anecdotes, which includes consumer complaints or testimonies on individual problems. Anecdotal evidence is problematic because it does not specify whether a particular problem is isolated or systemic or identify the economic source of the problem. Anecdotal evidence is helpful in the sense that it shows that at least someone is being harmed in some way, but it does not show how often a problem occurs or how severe that problem is. Without this specific data, it is hard justify a blanket solution that has real costs to taxpayers and consumers.[15]

One tool that agencies use to help make sure that regulations are as good as possible is benefit-cost analysis (BCA). If the expected benefits of a proposed regulation outweigh the expected costs, then it may make sense to issue that regulation. Ideally, measuring benefits and costs is helpful because it provides a way to tell the difference between a good regulation and a bad one. The techniques that the government uses to measure benefits and costs, however, are not always ideal. The analyses may be biased, or the results may not be appropriately interpreted.[16] In some cases, imperfect or biased BCAs may be worse than no analysis at all because decision makers justify their actions on faulty information.

Although the use of BCAs is growing, that may not necessarily be a good thing. In the worse cases, bureaucrats can use biased results strategically to impose regulations that they want to pass. Overstating the benefits and understating the costs make it easy to pass almost any regulation. Bureaucrats have the incentive to portray their regulations as having high benefits and few costs.[17] Further agencies cannot impartially judge their own rules and many analyses favor what agency leaders want, regardless of the true costs or benefits.

Beyond this direct problem of agency leadership's influence, BCAs can be problematic because it is difficult to accurately quantify the true costs and benefits of complex government action. Agencies use the value of a statistical life (VSL) models to pass many regulations. In the simplest terms, these models calculate a group's willingness to pay to avoid deadly risks. Agencies can then calculate how many lives their regulations are expected to save, and they multiply the VSL by how many lives will be saved. By multiplying the VSL by how many lives will be saved, agencies can estimate the total expected benefits from their proposed regulations.

VSL estimates are flawed for several reasons. First, each agency uses different data sets and comes up with widely varying numbers for each VSL. Even within agencies, VSL estimates can vary widely. For example, from 2005 to 2015 the Environmental Protection Agency used VSL estimates ranging from $6.8 million to $10.7 million for different rules.[18] Government researchers also do not choose consistent methodologies, which can further skew their estimates. Another problem is that agencies often use data subject to selection bias, further skewing the VSL estimates. When all of these factors combine, VSL calculations are usually inflated, which makes proposed regulations seem more beneficial than they actually will be.

When regulations are passed that actually have more costs than benefits, American taxpayers and consumers are harmed. Although regulations are meant to protect Americans, they restrict choice and increase costs. When companies have to pay the costs of complying with regulation, those costs get passed on to consumers. In some cases, regulations also limit privacy and restrict personal liberty.[19] Although regulations do often have significant benefits, they should be evaluated based on an objective comparison of their actual outcomes, not based on their intended benefits alone.

Making regulations is not flawless, and there are even government agencies that make sure regulations are made correctly and scientifically. The

Office of Information and Regulatory Affairs (OIRA) oversees how and why other agencies make regulations, but OIRA officials also work in a political world. OIRA officials are supposed to evaluate and criticize regulations, but there can often be a conflict of interest. The president's administration usually has overarching goals that officials in all agencies are trying to achieve, and OIRA officials do not necessarily want to sabotage their own administration's work.[20]

OIRA officials also run into another problem with the vice president. If OIRA officials wanted to block a regulation that another agency put forward, the agency proposing regulation can appeal to the vice president. OIRA officials do not want to create strife within the administration, so they only want to block regulations if they can win the political battle. If they know that the vice president is not likely to side with them, then OIRA officials have little incentive to block regulations knowing the vice president could unblock them.

The system of regulating, at both the state and federal levels, is problematic, and these problems tend to persist over time because it is difficult to get rid of old, redundant, or ineffective regulations. Especially at the federal level, regulations have continually accumulated over the years. For example, the Code of Federal Regulations consists of over 170,000 pages and over one million distinct restrictions. In the past 40 years, the federal code has more than doubled in size.[21]

Even though most regulations are intended to benefit the American people, the growing stack of regulations harms society in a few key ways. First heavy regulation hinders innovation and entrepreneurship, and second, overregulation makes overall economic performance less efficient. Limiting innovation, entrepreneurship, and efficiency slows economic progress and harms people, often more than the benefits that the regulations actually create.[22]

The harms of overregulation, especially as they reach the economy at large, are disproportionately borne by low-income households. When politics and public outcry get mixed into the regulation-making process, regulators can become hasty and motivated by a need to do something regardless of cost or outcome. Without deliberate research and careful considerations, regulations favor one group over another or fail to do what they were meant to do.[23] Once put in place, regulations can be difficult or nearly impossible to remove. Badly made regulations can exist on the books almost indefinitely, making life more complicated and costly for people who have to comply with them.[24]

Imperfect people with limited knowledge pass laws and regulations that everyone must comply with, or face the consequences. A critic might argue,

"Okay, the system isn't perfect, but is it fair to expect perfection from laws and regulations?" It is not fair to expect perfection, but it is fair to expect a realistic view about how policies are formed and what they can actually accomplish. Critics may argue that we simply need better, smarter policies to improve our electricity infrastructure. Better, smarter policies are nice in theory, but difficult in practice when we consider the incentives and constraints that policy makers face. Understanding how and why government officials actually make decisions can help taxpayers and voters more realistically evaluate the outcomes from electricity policies at both the federal and state level.

CHAPTER THREE

The Economics of Electricity

Good Intentions, Unexpected Outcomes

Even with the best intentions, government policies that are meant to help people can end up backfiring. Some policies do not achieve what they were intended to, and in the worst cases, policies can even end up doing more harm than good. Many states have enacted seat-belt laws that require drivers and passengers to buckle up or face a fine. Advocates for seat-belt laws say that seat belts save lives, which they do. Opponents of these laws argue that drivers and passengers have the right to decide if they want to wear a seat belt, even if it is a stupid decision. They claim that people should have the choice to decide whether they are willing to take the risk of not wearing a seat belt. Advocates might respond that public safety is more important than the liberty to choose not to wear your seat belt. Whichever side of the argument a person is on, seat-belt laws have undoubtedly had unintended consequences.

Economist Sam Peltzman studied the effects of seat-belt laws and found that overall traffic fatalities remained roughly the same in states that had passed seat-belt laws as in those that did not. Peltzman found that people buckled up more often in states with seat-belt laws, which makes sense because people want to avoid paying the fine for not complying with the law. Peltzman's most interesting finding was that mortality rates for people wearing seat belts in cars decreased, but this improvement was offset by an increase of deaths of pedestrians and cyclists.

Peltzman's findings were puzzling. Why would states where people buckle up more have the same rate of traffic fatalities as states where people tend to wear their seat belts less? It seems that seat-belt laws did exactly what they were supposed to do. People were putting on their seat belts more often. The problem was not necessarily the intentions or the contents of any of the seat-belt laws; the problem was how people responded to the law in ways that lawmakers could not predict. Once required by law to wear a seat belt, people wore their seat belts more often. But because the seat belts made them feel safer, they drove more aggressively, resulting in more crashes. Thus states that passed seat-belt laws saw higher rates of people wearing their seat belts, but still had the same rate of people dying in car accidents.

On average, drivers in states with a seat-belt law had a lower probability of dying in a car crash. Unexpectedly, states with these laws experienced higher rates of traffic accidents. When wearing a seat belt, the average driver feels like he or she can drive more aggressively, recklessly, or distractedly. No rational driver wants to get in a wreck, but when the population as a whole drives just slightly worse, more accidents happen. When looking just at the death rates of drivers, the law seemed to work because the rate went down slightly. Even though drivers were safer because of the seat belts, everyone was in slightly more danger. Drivers ended up hitting and killing pedestrians and cyclists at a higher rate. Seat-belt laws may have saved more lives of drivers, but pedestrians and cyclists suffered increased risks under the new law.[1]

Lawmakers failed to foresee how people would actually react. Unfortunately, it is impossible for policy makers to know exactly what all of the effects of their laws will be. If government officials had perfect knowledge and perfect foresight, then maybe laws could work out exactly as they are meant to. The lawmaking processes are messy and complicated. One of the problems with making laws is that there are millions of individual people affected by those laws who respond to them in ways that are not easily predictable. No one person, no matter how smart or educated, has enough knowledge to foresee what all of the consequences of a law will be, both intended and unintended. And because we have created laws that affect nearly every aspect of our everyday lives, it is no surprise that these overlapping laws often create negative consequences that each and every one of us has to live with.

Should we have no laws because the effects are not perfectly predictable? The thought of anarchy does not set well with most people. Understanding the practical limitations of the policy-making process is important for both policy makers and members of the public. Policy makers can make smarter policies when they better understand how unintended consequences happen. Policy makers would do well to be more forward thinking when proposing new laws and regulations.

Using the Economic Way of Thinking

At its most basic level economics is the study of how people make decisions and exchange with one another. Economics helps us understand why people do what they do. This understanding can help us make better informed decisions in all areas of life, not just regarding money and profits.

One of the assumptions that underlays economics is the notion that individuals want to take actions that will make them better off, and that one of the most effective ways of doing so is trading with other people. When people are free to exchange as they see fit, people are more likely to get what they want, or at least to make their lives better than they would have been if they could not trade with other people. Economists often refer to trade of this sort as mutually beneficial exchange, because if an exchange does not benefit both people involved, the exchange does not happen. One person in the exchange will choose not to participate if he or she knows that trading would make him or her worse off. Mutually beneficial exchange is central to economics as a discipline and to understanding the economic way of thinking.

One of the easiest examples of mutually beneficial exchange comes from the following scenario that illustrates bartering. One farmer grows potatoes, leaving little time to raise dairy cows for milk. But a dairy farmer next door raises milk cows. Our potato farmer can approach the dairy farmer and suggest an exchange. If the potato farmer and the dairy farmer come to an agreement that makes both of them better off, then they will exchange goods with one another. They may decide that one gallon of milk is worth a pound of potatoes, or they may decide that a gallon of milk is worth ten pounds of potatoes. It is impossible to say how much the two farmers will exchange because they both have preferences that change over time and in different circumstances. One thing is certain though. If the two farmers are free to choose for themselves, they will exchange only when the trade is beneficial for both of them. The two farmers have created a market, exchanging things to make themselves better off.

This notion of mutually beneficial exchange is a useful way to begin to think about electricity. People want access to reliable electricity, and they also have other preferences. In the late 1800s and early 1900s, entrepreneurs set up generators and constructed transmission lines for people who were willing to pay for electricity. One of the key facilitators of this exchange is money. Humans do not have to barter with potatoes and milk anymore because money takes the place of bartering. Money is a substitute that we use so that we can exchange anything we want at any time. An electricity market emerged where paying customers and electricity producers engaged in mutually beneficial exchange. Today, the electricity market has become more complex, but the basic principles remain the same.

Money easily tells us the price of something, and prices are important because they communicate information. Ask any child what a price is, and he or she will tell you that a price is how much something is worth. But what does "worth" mean, and how is it determined?

Adam Smith, the father of modern economics, was puzzled by a paradox over how much worth we assign to water and diamonds. Water is essential for human life. We need it to drink, cook, clean, and grow crops. Water is fundamentally necessary to human life beyond almost anything else, but despite how essential it is to human life, it tends to be relatively inexpensive. Small amounts of diamonds, on the other hand, are worth thousands of dollars. Diamonds are not necessary to life by any means—they do not provide for any of our basic human needs. Diamonds are little more than a sparkly rock. A price, whether for diamonds or water, communicates how much people value something and how scarce that thing is. Prices also communicate the value of all the resources that were necessary to produce a good or service. Diamonds are highly valued, costly to mine, and relatively scarce, which makes them expensive. Water, on the other hand, is highly valued, but it is so abundant and relatively easy to produce that residential water rates cost a fraction of a penny for a gallon. Prices are critical to understanding what people value and what is scarce.

Prices can easily change when preferences and circumstances change. For example, in the average American home, a glass of tap water costs a small fraction of a penny. The cost of a single glass of water is so small that nearly nobody ever thinks about it. Imagine, however, a different scenario. A traveler is lost in the Sahara and has not had any water for a whole day. As he stumbles around the desert, he sees an oasis in the distance. As he runs toward the oasis, he sees that the oasis has only a few trees and there is a woman standing beside a well. She is charging $100 for each glass of the well's water. The lost man has a $100 bill in his wallet. What happens? We all clearly recognize that the man gladly hands over the $100 for a glass of water. Why would he pay a fraction of a penny of water for a glass of water at home but $100 at the oasis? He was in a different place under different circumstances. Water had become scarce to him, and his subjective valuation of water had changed.

Government policies are particularly effective at distorting mutually beneficial exchange and skewing price signals. Without correct price signals, we cannot communicate necessary information accurately. For example, when a natural disaster occurs, prices communicate the scarcity of certain goods so that those goods can be used in their highest valued uses.

In 1996, Hurricane Fran devastated parts of North Carolina, blocking roads and causing blackouts. The weather was hot and muggy, and people needed ice to cool necessities like insulin and baby formula. Four young men from an area of the state that was not affected by the hurricane went to dozens of stores and bought up as many bags of ice as they could. These men bought the ice for $1.70 a bag, but when they arrived at the areas affected by the hurricane, they began selling bags for over $8 a bag. Although people had lined up and were willing to pay these high prices, local police soon showed up and arrested the young men for violating the state's price gouging laws. According to the law, it was illegal to charge "too much" for ice or any other good, even in times of emergency.[2]

Because a government law prevented prices from fluctuating to communicate scarcity, people in the aftermath of the hurricane were left without ice to cool their medications or keep their babies' food from going bad. If prices were kept low, then potential entrepreneurs would not have any incentive to resell ice to hurricane victims. And even if people had been willing to sell ice to hurricane victims at a lower price, there would have most likely been a shortage, and the ice would not have gone to the people who needed it the most. When prices go up because a good becomes scarce, people who really need ice, for keeping medications cooled for example, would be willing to pay a higher price than those who would have liked to have ice just as a convenience. When prices for scarce goods are allowed to fluctuate, those goods are able to go to their most highly valued uses. In the case of Hurricane Fran, price gouging laws stopped prices from communicating information like they are supposed to. As a result, people could not get the things that they desperately needed.

Even with good intentions, government policies can distort the economics of electricity because that is precisely what they are designed to do. Subsidies, for example, are purposefully designed to make expensive things cheaper. Loan guarantees are meant to make risky investments less risky. Mandates are meant to force people to do things that they would not have otherwise done. All of these policies are ways that government intervenes in markets, distorts prices, and spurs unintended consequences.

Competitive markets without much government intervention generally produce better outcomes for both buyers and sellers. This principle applies to almost any industry. More competitive markets can function better when there are not many barriers to entering or exiting the market.[3] When government policies make it harder for new companies, ideas, or technologies to enter an industry, markets become artificially restricted, and society as a whole may be losing out on cheaper prices or better goods and services. When government policies give special privileges to handpicked industries or companies, the benefits from exchange fall because competition falls, too.[4] One example of this is the taxi industry versus ride-sharing companies like Uber and Lyft. In most cities, taxis are required to purchase government-issued permits or medallions to operate. This requirement effectively limits competition in the market and favors those people who are lucky enough or rich enough to get the permits. Ride-sharing companies have circumvented the monopoly that taxis have, and now offer services consumers want generally at a lower price and with increased convenience. Due to the new competition, some taxi companies are going to the municipal governments to make the laws more strict, effectively attempting to legislate away their competition, leaving consumers worse off because of their political lobbying.

The Seen and the Unseen

Some policies may benefit large groups of people at the expense of a small group, and other policies benefit small groups of people at the expense of many. Even when politicians and bureaucrats want to help out the majority at the expense of a minority group, that outcome may not happen. Government officials, just like anyone else, are bad at foreseeing the future. When policies change the ways that large groups of people make complex decisions, unexpected things happen. In the worst cases, some government policies can completely backfire when they produce an outcome that is exactly the opposite of what they were meant to achieve.

Frederic Bastiat, a French philosopher and economist in the 19th century, wrote about the danger of assuming that government officials know enough to make perfect decisions. Bastiat talked about the phenomenon of the "seen" and the "unseen."[5] When government officials create a policy with a particular goal in mind, they often expect a certain outcome. This outcome is the "seen" one. For example, in the 1970s, Congress passed the Endangered Species Act to protect species that are near extinction. Under this law, it is illegal to damage the habitat of one of these species or to kill, capture, or otherwise harm an individual of a protected species. By protecting individual organisms and their habitat, we are saving species, right? It seems obvious that outlawing the killing of species and prohibiting the destruction of their habitat would prevent them from going extinct. This example is the "seen" effect of the law.

Bastiat, however, warned of the "unseen" or indirect effects of laws. The Endangered Species Act has spurred unintended consequences that policy makers were unable to foresee. Under the provisions of the Endangered Species Act, people cannot use their own private land when an endangered species is located there. When a landowner finds an endangered species on his land, he or she may fear losing the right to use the property as he or she intended. Endangered species can destroy the livelihoods of farmers and ranchers, because once an endangered species is present, they cannot work the land they are dependent on. The Endangered Species Act lacks any sort of compensation mechanism when an endangered species is located on privately owned land. Often, landowners then engage in what is termed "shoot, shovel, and shut up." To make sure their livelihoods are not destroyed, landowners may kill endangered species that threaten that ability. They then "shovel" by covering up evidence that the species ever lived there. Finally, they will tell no one (especially a government official whose job it is to protect the species) that an animal was there for fear of punishment. This outcome is what Bastiat would call the "unseen." Even with the best intentions, policy makers have a difficult time foreseeing what policies will actually do.

Nobel Prize–winning economist Friedrich Hayek further illuminated Bastiat's ideas about the nature of knowledge. Hayek wrote of a "knowledge problem" that occurs when anyone tries to plan a complex system. Even the most educated advisers using the most sophisticated models do not have enough information to accurately or adequately plan out complex systems.[6] The knowledge problem happens because a central planner cannot know what is known by the system as a whole. Government officials try to plan complex systems every day, and many of the negative outcomes of this policy-oriented planning happens because of the knowledge problem. Policy makers work in a world where millions of individuals are making choices for themselves. When policy makers create a law or regulation to do one specific thing, the millions of individuals who are affected by the law choose to respond in uniquely individual ways. Policy makers have no way to know what the exact outcomes of their policies will be because they do not have access to the local knowledge distributed throughout the economy.

Hayek described two types of planning: centralized and decentralized. Government officials who centrally plan try to create policies that will spark a specific outcome, but almost invariably, unintended side effects result. Decentralized planning harnesses the power and knowledge of millions of people. Decentralized planning takes place in markets. People work together to exchange goods and services that benefit them, but they also exchange knowledge that benefits them, too.

Hayek tells a story of tin manufacturing to show how decentralized planning works. Imagine that the world's largest tin mine was suddenly closed and could not produce any more. If tin users continued to consume tin at their regular rates, they would exhaust the supply and there would be a shortage. But this shortage is not likely to happen in a functioning market because changes in the price of tin will reflect the new scarcity. Prices become a tool to communicate knowledge about the scarcity of tin. When tin becomes more expensive, tin users have to make decisions about how much tin is worth to them. Tin will then be allocated to its more highly valued uses when the people who want it most pay the most for it. Once the price is raised to reflect the scarcity, some of the tin can be more profitably employed elsewhere. People choose to economize on tin when the price is higher, avoiding a shortage. In this case, no central planner was necessary to ration or allocate the tin. The collective workings of the market made sure that tin was still available for its most highly valued uses. Tin users did not need to know the details about the closure of the mine or exactly how much tin was still available. The new price of tin communicated the knowledge necessary to avoid a shortage. People would then begin to cope with the higher cost of tin by opening up new tin mines or finding substitutes, like aluminum.

In theory, if a central planner had access to all of the information dispersed throughout the economy, the central planner could only do as well as the decentralized system, but not better. Since it is impossible for a central planner to have all of the information dispersed in an economy, centralized planning will necessarily be problematic. The collective brain of millions of people working together produces outcomes that more accurately reflect what people want than a government official sitting in a cubicle in Washington, D.C. Just like any policy area, a central planner cannot possibly gather all the information he or she would need to plan out America's electricity market. When government policies attempt to do so, unintended consequences result. The reality of the importance of dispersed knowledge has more recently been described by James Surowiecki, in his book *The Wisdom of Crowds*, which details the latent knowledge that can be harnessed from the wider population.[7]

The Economics of Government Favoritism

Government officials distort economic functions by giving special interest groups privileges. When special interest groups lobby for these privileges, elected officials often oblige by extending corporate bailouts, subsidies, price and entry regulations, tax credits, contrived monopolies, noncompetitive bids, loan guarantees, and trade protections.[8] Politicians are usually experts in some field, but they cannot be experts in every field. Because very few members of Congress are experts in energy production or environmental science, they are prone to support energy corporations that are best at lobbying and who give the most financial support.[9] The effects of government favoritism are wide and sweeping. These policies restrict market competition, often leading to higher prices and lower quality goods and services. When politicians codify special privileges into law, society as a whole bears the costs while the privileged few reap the benefits.

Government officials choose winners and losers, and nobody wants to be the loser. Special interest groups or companies try to persuade politicians and bureaucrats to pick them as the winners while also choosing their competitors as losers. The process of trying to get government officials to pick a winner is called "rent seeking." Many interest groups eagerly invest time and money into rent seeking so that they can gain or maintain the status of winner. They donate to political campaigns and political action committees in a kind of favor-trading service. They can also advertise for public policies that benefit them, hoping that regular citizens are persuaded, and influence policy makers on behalf of the interest group. The most direct way to get chosen as a winner is to heavily lobby policy makers.[10]

The negative impacts of government favoritism do not stop here. The complete waste of time, money, and resources used for lobbying are particularly stark. Industries or companies that want the government's good favor invest time and money that could have been put to better social use. Instead of spending time and money persuading politicians for privileges, these companies could have been engaging in research to find more efficient, cost-effective, or reliable products and services—these losses are commonly called "dead-weight losses" by economists.[11]

When government distorts markets, it limits how people can engage in mutually beneficial exchange. In the worst cases, government-granted privilege makes rich, well-connected people even richer while everyone else bears the cost. For people who are not in the ruling elite, preferential treatment generally causes prices to go up, quality to go down, and innovation to stagnate.[12] Government favoritism also causes a net loss to the economy overall. The benefit to a single industry or interest will inevitably be less than the total damage to the rest of the community.[13] Some of the most destructive forms of privilege are subsidies, loan guarantees, and restricted market entry.

Government policies and programs can unintentionally cause moral hazard, which is a phenomenon that happens when people are protected from the effects of their risky decisions. When people feel safer, they act in riskier ways because they feel protected against harm or losses.[14] Seatbelt laws are a classic example of how government policies can introduce moral hazard.

In the electricity industry, moral hazard can take many forms. For example, under a loan guarantee program, the government agrees to foot the bill if a company fails. If the company is successful, it gets to keep the profits, but if it is not, then taxpayers cover the costs of the failure. Loan guarantees distort the economic considerations that a company would normally make, and so companies make riskier investments that they would not make if they had to cover the costs of failure themselves.[15]

A side effect of loan guarantees is that companies, in an effort to secure a loan backed by taxpayers, focus on pleasing government officials rather than their customers. Producers shift their resources away from what consumers want and devote them to pleasing government officials. Government-approved projects may not have the same merit or viability that projects would have if they were based on supply and demand in the market. This shift in decision making can lead to a misallocation of resources because industry leaders make decisions based on contradictory information from the government and the market.

A large number of loan guarantees is a sign of cronyism. Politicians and bureaucrats use loan guarantees to pick winners and losers in a particular industry. Federal loan guarantees most often go to large corporations, not to small start-ups. Some of these large corporations double-dip with government money because they receive other government grants. In one of the best examples of this behavior, the federal government granted St. Clair Solar $192.9 million in two loans to buy solar panels from First Solar, but the problem was that St. Clair Solar is a wholly owned subsidiary of First Solar. In essence, St. Clair Solar received loan guarantees and grants to help it buy solar panels from itself.[16]

One of the worst and most notorious loan guarantee failures happened with an energy company called Solyndra. In 2009, the federal government agreed to give Solyndra a $535 million loan guarantee. After two years, Solyndra filed for bankruptcy and fired its 1,100 employees, sticking taxpayers with the cost of the loan.[17] More recently, SunEdison was another major energy corporation that failed after it had received hundreds of millions of taxpayer dollars. SunEdison's total support from all government-granted loans, loan guarantees, tax-exempt bonds, insurance, and subsidies was roughly $1.5 billion.[18]

Further negative impacts come from the large numbers of complex regulations as they discourage entrepreneurship and innovation from happening.[19] When people have to worry about complying with regulations, productivity goes down because they have to devote their scarce time and money to making sure they are compliant.[20] As the list of regulatory requirements continues to grow, business owners and government officials must spend increasing amounts of time and money on compliance, rather than benefiting themselves and society through innovation, value creation, or entrepreneurship.

There are trade-offs for companies who choose to allocate time and money to gain government privileges. Corporations who receive government-granted privileges are less likely to innovate and thus less likely to discover better goods and services.[21] As companies put more time and effort into getting special privileges, they invest fewer resources in research and development that could bring about new innovations.[22] If a company is devoting more resources to lobby the government, those resources are not available to put toward creating new technology or additional value for a company's customers. Companies that lobby the most are often older, less diversified, and less profitable than companies that do not lobby.[23] Larger, more established companies often have more resources to devote to lobbying the government than smaller start-ups. When government picks winners and losers through policy, making friends in high places becomes a substitute for innovation and entrepreneurship. Competition allows innovative companies to challenge less innovative ones, and with the incentive to make a profit, competition benefits everyone because companies are developing new technologies that make life better.[24]

Subsidies are one of the most blatant forms of government-granted privilege because they are direct handouts of taxpayer money to businesses that politicians and bureaucrats have cherry-picked. Almost every industry imaginable receives subsidies, and the electricity sector is no exception. Laws like the Energy Policy Act of 2005 provide billions of dollars of subsidies to nearly every industry in the energy sector.[25] However, when markets are free, potential competitors can exchange with willing buyers, and prices go down as sellers try to compete with one another for more customers. Society is worse off when markets are restricted because there is less competition, fewer choices, and less total value created.[26]

CHAPTER FOUR

Wind Power

Wind Power: Not in My Backyard

One of the biggest winners of modern energy policy is wind power. In 2013 wind power received $5.9 billion in federal subsidies—more than any other energy source.[1] But despite this favorable treatment, wind power often faces opposition from those who do not want to live with a wind turbine in their backyard. It is hard to blame them for opposing wind farms in their neighborhoods. It is not surprising that people might oppose the construction of towers with spinning blades that are hundreds of feet tall near their homes. Many people consider wind turbines eyesores that can block otherwise beautiful views and reduce property values. This local opposition is often called NIMBY-ism, or the "not in my backyard" mentality.

Though it's not surprising that people may not want to live in the shadow of a massive metal tower, NIMBY-ism often creates a surprising conflict between groups of environmentalists. This clash most conspicuously played out in America's millionaire playground, Cape Cod. The Kennedys, America's most famous political family, had been fervent supporters of wind power and other renewable energy sources for decades, but when a wind farm was proposed off the shore of their Cape Cod mansions, they caught a bad case of NIMBY-ism.

The Kennedys are not the only ones with NIMBY-ism. It happens all across the country. Proponents of wind energy often extol wind power's environmental benefits when they justify funding it with taxpayer dollars and mandating it through state laws. But despite these perceived environmental benefits, it is difficult to find people who want giant wind turbines installed near their neighborhoods. If wind turbines are so environmentally friendly, then why are people so hesitant to live near them? The answer to this question can be found by examining who benefits from wind power, and who bears the costs. It is easy to support the construction of new wind power capacity when you do not have to deal with the unpleasant consequences of living in the shadow of a wind turbine.

The Kennedys were not the only ones to oppose a wind farm in Cape Cod. Fishers and Native American groups in New England agreed with the Kennedy family, opposing the wind farm off Nantucket Sound. After years of political fighting and litigation, the Cape Wind project died.

Not only did politics and lawsuits bring down Cape Wind, but the final death blow to the project was the company's own poor management. In January 2015, two of the biggest utilities in Massachusetts terminated their contracts with Cape Wind because the project was not meeting its agreements. Northeast Utilities and National Grid had previously agreed to buy power from the proposed wind farm through a power purchase agreement (PPA), but Cape Wind failed to meet the deadlines specified in the contract. PPAs are agreements between utilities and independent energy producers that provide certainty by guaranteeing a fixed rate for electricity produced for a certain time period.

Caroline Pretyman, the spokesperson for Northeast Utilities, stated, "Unfortunately, Cape Wind has missed these critical milestones. Additionally, Cape Wind has chosen not to exercise their right to post financial security in order to extend the contract deadlines. Therefore, the contract is now terminated." Jake Navarro, the spokesperson for National Grid, similarly stated that the utility provider was "disappointed that Cape Wind has been unable to meet its commitments under the contract, resulting in today's termination of the power purchase agreement."[2]

The proposed Cape Wind project was also expected to have high costs for electricity consumers in the state. Associated Industries of Massachusetts, the largest employer association in Massachusetts, spoke out against the project because of its expected impacts on ratepayers. John Regan, spokesman for this employer association, said,

> Associated Industries of Massachusetts and its 4,500 member employers strongly endorse the decisions by Northeast Utilities and National Grid to terminate their long-term contracts with Cape Wind. These power companies deserve tremendous credit for taking actions that will save customers billions of dollars that would otherwise have been spent buying electricity at the highest power price ever negotiated in Massachusetts.

Regan also said that the organization supports renewable power projects when they are economical. In this case, as Regan put it, the "staggering costs" of Cape Wind "would have saddled ratepayers for decades to come.[3]

Back in 2009, the state and local governments issued permits to Cape Wind, as long as the company constructed the transmission lines before May 2015. The 2015 deadline proved unrealistic due to the lawsuits, public opinion, and management issues. Cape Wind wanted to extend its permits until May 2017, so it approached the Energy Facilities Siting Board to gain that extension. The purpose of the board is to ensure that electricity is produced at the lowest possible cost and with the smallest environmental impact. In April 2016, the board found that Cape Wind did not meet their requirements and unanimously rejected Cape Wind's petition for an extension. Cape Wind appealed the board's decision to the Massachusetts Supreme Judicial Court, but withdrew its appeal in October. After spending tens of millions of dollars over a decade in a losing battle, it is highly unlikely that Cape Wind will be completed.

The fight over wind farms in New England did not end with the demise of the Cape Cod project. In 2016, Deepwater Wind LLC entered the final construction phase of Block Island Wind Farm off the coast of Rhode Island. When completed, the 589-foot-tall turbines of Block Island will be the first offshore wind power plant in the United States. Like the Cape Cod project, the Block Island Wind Farm has also been controversial. Opponents of the project feared that the towering turbines would lower the value of their Block Island properties. They also feared that the wind farm would lower rental incomes in the area and harm public resources like the Block Island Town Beach. Attorney Terence Tierney explained, "The Rhode Island Coastal Resources Management Council is rubber-stamping this as fast as they can. This is not an open and transparent process. It is a process that excludes virtually everyone but [Deepwater Wind]."[4]

Because of these concerns, in 2013, five Rhode Island residents asked the state's Superior Court for an injunction to force the Rhode Island Coastal Resources Management Council (CRMC) to allow locals to participate in the review of the proposed wind farm. The CRMC's purpose is to issue permits for construction in the state's coastal zone. The residents were successful, and the CRMC held several public hearings where proponents and opponents of the project were able to voice their views. The controversy continued in Rhode Island for several years until the project was given the go-ahead by state agencies.[5]

The opponents of the Block Island project were not nearly as successful in their NIMBY fight as the Kennedy family. The federal government has approved many more locations along the East Coast for offshore wind projects. Under Section 388 of the Energy Policy Act of 2005, the secretary of interior can grant leases for offshore wind projects, which he delegated to the director of the Bureau of Ocean Energy Management.[6] As these projects continue, we can expect more NIMBY fights to arise.

Wind as the Federal Government's Winner

Climate change and pollution are two problems that worry many environmentally conscious Americans. Neither problem, however, has a simple solution. Despite the complexity of the issues, renewable energy is often cast as a one-size-fits-all solution. Many claim that if we use more renewables and less fossil fuels, we can solve the problem of climate change and prevent future catastrophe. Politicians and bureaucrats in the government rarely have the best solutions. Even with the best of intentions, policy makers have limited information about the impacts of a proposed policy and whether it will achieve its goal. Government officials pick winners and losers with their policies, and for the past few decades, wind power has been the winner in the energy sector.

Why has wind energy become the nation's most heavily subsidized energy source? First, wind power is politically popular. The sight of giant wind turbines is impossible to miss, and by supporting a conspicuous energy source, government officials can take credit for promoting "green" policies. Second, the powerful wind lobby continues to seek special privileges from the government that allow them to remain winners in the energy sector year after year.

Many people may say, "Does it really matter if the government has chosen wind power as the winner? Isn't it a good thing that we support a carbonless form of energy with tax dollars?" On the surface, wind may seem like a more prudent investment of tax dollars than other energy sources because it does not emit carbon or pollution when it generates electricity. Wind power, however, cannot easily or cheaply solve the environmental ills of our electricity sector. Wind power has economic implications that are often hidden from the average electricity consumer.

The economics of wind power are not as straightforward as most people assume. On the surface, there are obvious costs that are easy to see. These costs include the manufacturing of the turbines, operations and maintenance costs, and decommissioning costs. But the unseen costs are equal to or more important than the seen ones. Government policies, such as subsidies and mandates, distort energy markets, which often raise prices for consumers and raise taxes for taxpayers. When the seen and the unseen costs are combined, energy consumers have higher electricity bills and pay more in taxes to support wind power than they realize.

The recent surge in the wind industry is not the result of wind power's ability to provide reliable and affordable energy to consumers. In 2002, wind produced just over 10 gigawatt-hours of electricity, but by 2015, wind produced 191 gigawatt-hours.[7] Rather than market forces, the main drive for the wind industry's growth has been government support. All levels of government, including the federal, state, and local levels, have been subsidizing and mandating wind power as early as the 1970s. The most robust subsidies and mandates, however, were enacted in the 1990s and early 2000s. These policies have given the industry a huge advantage over other sources of electricity. Without the government's help, the whole wind industry stumbles.

The federal government's largest handout to the wind industry is the production tax credit (PTC). The PTC was initially set up in 1992 to be a temporary form of "training wheels" to get the wind industry going. Two and a half decades later, the PTC is still rolling along, despite the fact that it has expired and been renewed several times. When the PTC has expired in the past, investments in wind power drop. The PTC expired in 2000, 2002, 2004, 2010, and 2013. After each expiration, new wind installations plummeted between 70 and 92 percent.[8] Congress reauthorized the PTC after each expiration, allowing the amount of new wind installations to grow again until a new expiration when the investments would once again plummet. After the PTC expired in 2013, Congress added a provision to the American Taxpayer Relief Act, which allowed projects that began construction before 2014 to receive the tax credit.[9] The Tax Increase Prevention Act of 2014 was yet another training wheel, extending the PTC so that any project that began work before January 1, 2015, was eligible for the credit.[10]

The PTC allows private investors to profit at the taxpayers' expense. Warren Buffett, one of the richest people in the world and a wind industry magnate, has noted that "[O]n wind energy, we get a tax credit if we build a lot of wind farms. That's the only reason to build them. They don't make sense without the tax credit."[11] Without government assistance like the PTC, wind power is an unappealing investment. Policies like the PTC essentially guarantee that wind power producers can make profits. When subsidies are taken away, the huge drop in wind investments suggests that wind power is not as economically sustainable as advocates claim.

In addition to the PTC, Congress has created 82 separate incentives or initiatives to promote wind power. Nine different federal agencies oversee each of these programs, which provide wind power financial support.[12]

Each one of these policies creates hidden impacts that will ultimately have to be paid for by American taxpayers and consumers.

What Happens When the Wind Stops Blowing?

Besides its complicated economics, wind power also has physical weaknesses. When compared to other sources of electricity, wind power is inconsistent and intermittent because the wind does not always blow when people want to use electricity. For decades, coal and natural gas have been our major sources of electricity largely because they are dispatchable. Dispatchable energy sources can be turned on or off, or ramped up and down, to meet fluctuations in electricity demand. Their fuel sources can also be transported to where they are most needed. Intermittent power sources like wind are not dispatchable because we cannot control when the wind blows and we are not able to store wind power to be used later. We can control, however, when we burn more coal or natural gas. Wind's intermittency makes it one of the least reliable forms of electricity generation.

One potential solution to wind's intermittency problem is grid-scale storage. For decades, innovators and entrepreneurs have been searching for ways to store electricity on a massive scale. Conventional batteries are incredibly costly and have undeniable environmental impacts. Alternative forms of electricity storage could help overcome wind power's biggest flaw, but grid-scale storage does not seem to be viable in the foreseeable future. Until the day when we have large-scale storage, wind power and other nondispatchable sources of energy like solar will continue to suffer from reliability issues.

Another of wind power's shortcomings is that it cannot currently provide baseload power. Although electricity demand fluctuates throughout the day, there is, more or less, a base level of electricity that power plants must provide. Wind's intermittency makes it nearly impossible for wind farms to provide the consistent base of power that Americans need every day.

Some wind advocates argue that connecting multiple wind farms across a broad region can help to reduce wind power's reliability shortcomings. Advocates argue that even if the wind slows or stops in one place, the wind is most likely blowing in another, so interconnecting wind farms overcomes more of the inconsistency problem. Researchers from Carnegie Mellon University studied how interconnections could help make wind power more reliable. Using 20 wind plants in Texas, the researchers found that interconnecting four separate wind plants can reduce output variability by up to 87 percent. Connecting more and more wind plants, however, did not continue to decrease the variability. Connecting 16 more wind plants to the original 4 only reduced variability by an additional 8 percent.[13]

Overall, scientists disagree on how much interconnecting wind farms can make wind power more reliable. Of course, the effectiveness of these interconnections will be unique to each location's specific characteristics. Interconnection infrastructure also comes with a price. Installing continually growing amounts of interconnection infrastructure will cost wind power producers more money, and those costs will be passed along to energy consumers in the form of higher electricity bills and to taxpayers from ever-growing government assistance to the wind industry.

As government subsidies cause nondispatchable sources of electricity to become more widespread, the electrical grid as a whole becomes more and more difficult to control. Wind's variability means that we need to have backup sources of power when there is not enough electricity to meet demand or when the wind unexpectedly stops blowing in a certain area. When the opposite occurs, and wind produces more power than is needed, we have to disconnect wind farms from the grid. In some cases, wind producers actually pay utility companies to accept wind-generated electricity when more of it is being produced than is demanded. The electrical grid was not designed to cope with variable generation sources that we cannot necessarily control, so grid managers have to be creative when dealing with the inconsistent nature of wind and other nondispatchable sources of energy.

Providing backup sources of electricity is important if an area is heavily dependent on wind power. If the wind suddenly stops, people do not stop using electricity. Grid managers use existing power plants to generate electricity when wind farms do not produce enough or unexpectedly decrease their output. These backup sources of electricity are called "operating reserves," which are divided into two categories: cycling and noncycling. Cycling reserves are power plants that can change their output quickly to meet sudden fluctuations in energy supply or demand. Cycling reserves are continually producing electricity all day, but they are kept running at a lower level than their full potential. Whenever electricity supply or demand changes, cycling reserves can change their outputs easily and quickly to keep the amount of electricity in the grid stable and consistent. Two of the best forms of cycling reserves are hydropower and natural gas because they can increase their output quickly. Dams only need to let out a little extra water through their turbines, and natural gas plants only need to burn extra natural gas to spin their turbines.

Noncycling reserves are not meant to be ramped up and down as quickly as cycling reserves. Throughout the day, there are predictable patterns of energy consumption that grid managers can plan for. In most places, energy demand peaks in the morning as people are getting ready to go to work or school. A second peak happens in the evening when people return home.

These general patterns are so regular that grid managers keep certain power plants offline until they are needed. Noncycling plants require between 10 and 30 minutes to get up and running to compensate for the predictable changes in energy demand throughout the day.

Because wind can be difficult to predict, wind power requires cycling reserves to be ready for unexpected changes in supply. As government subsidies and mandates make the wind industry grow artificially fast, more cycling reserves will be necessary. Frequently ramping cycling reserves up and down often results in higher operations and maintenance costs for traditional power plants. Most fossil fuel plants are designed to provide a steady baseload of power, so ramping them up and down more often causes wear and tear and often makes them less efficient. Fossil fuel plants that are used as cycling reserves for wind often emit more pollution and require more maintenance per unit of energy they produce because of the added stress of intermittent renewables like wind.[14]

Sometimes wind power produces too much electricity rather than not enough. Baseload power sources, like coal, nuclear, and hydropower, produce a fairly constant output of electricity. Occasionally, a sudden burst of wind may cause a wind farm to produce excess power, which can put too much electricity into the grid and threaten its stability. Baseload power sources can be difficult to ramp up and down quickly, and disconnecting them could throw the whole grid into chaos because they produce the majority of the electricity. Grid managers in some situations must disconnect entire wind farms from the grid to preserve their stability. The electricity that a wind farm produces in these circumstances is wasted because there is no way to effectively store it. Wind power producers, however, receive PTC subsidies regardless of whether anyone actually uses the electricity they produce, which also creates an economic waste.

In other cases, when wind farms produce too much electricity, grid managers do not necessarily have to disconnect them from the grid. When electricity demand is lowest at night, the wind blows hardest in many locations, meaning that wind-generated electricity is produced at the highest levels when electricity is least needed. If more wind-generated electricity is produced that is demanded, wind power producers get rid of the excess electricity by paying utility companies to take it. This phenomenon is called "negative pricing."

Under the PTC, wind producers get paid $23 for every megawatt-hour they produce, no matter what market factors are present. Wind power producers can then pay utilities to take their electricity while still making a profit, or, at the very least, breaking even.[15] The PTC's subsidies distort the energy market so that wind producers are immune from the normal forces of supply and demand. In an unsubsidized world, wind power producers would have to bear the cost of paying utility companies to take their electricity, rather than utilities willingly buying it to meet demand. Taxpayers are then responsible for footing the bill for wind power producers who must sell their excess electricity.

Geography is another main drawback of wind power. Wind power can only be generated efficiently in windy places, but the windiest places are usually far from population centers. To effectively use wind power, we have to go to the windy places, build wind turbines, and then build transmission infrastructure to ship the electricity where people want to consume it. All of these steps add costs and create challenges for providing reliable and affordable electricity.

Fossil fuels and nuclear power do not face this same challenge, as they are dispatchable sources of power. We mine the fossil fuels and uranium needed to produce electricity and ship them to power plants. Those fuel sources can be stored until they are needed to generate electricity at power plants, which are usually located relatively close to population centers. The highest concentrations of onshore wind are usually remotely located. The Great Plains in Montana, Wyoming, and parts of the Midwest are some the windiest places in the country, but they also have some of the smallest populations. Potential entrepreneurs in the wind industry could take advantage of these windy places, but building wind farms in the windiest states means that wind power producers must build transmission infrastructure for regional grids where people want to use the electricity. Entrepreneurs in the wind industry are already facing the harsh reality of wind's geography problems. Gabriel Alonso, who works as the chief development officer of Horizon Wind Energy, said, "The windiest sites have not been built because there is no way to move that electricity from there to the load centers."[16]

Turbines on the wind-swept plains of Wyoming could produce roughly 50 percent more electricity than the same turbines in New York, but the fact that Wyoming is so remote makes taking advantage of all the moving air extremely costly. When producers incur higher costs, they pass those costs along to energy consumers. If the government offsets those higher costs with subsidies, the price of wind energy does not miraculously become cheaper. The true cost of wind power becomes more difficult to see because we pay for government assistance through our taxes. One way or another, all of the costs of generating electricity must be paid for, either by energy consumers or by taxpayers.

Determining how much new infrastructure will raise electricity bills or taxes is difficult because each circumstance is different. For example, the

Public Utility Commission of Texas (PUC) contracted private companies to build 3,589 miles of new transmission infrastructure for wind power from 2005 to 2013. The contracted companies spent $6.8 billion to connect isolated wind farms to urban areas.[17] To cover these costs, Texas utility companies are forced to charge customers higher rates. The PUC estimated that the transmission project would add $3 to $5 to each consumer's monthly electricity bills for the next 10 years. These rate estimates were made when the project was expected to cost $4.9 billion.[18] Because the project's actual cost was nearly $2 billion higher than expected, Texans' electricity bills will likely be even higher than the PUC estimated.

Although most wind turbines are located on land, some renewable energy advocates are calling for an increase in the amount of offshore wind power production. Most coasts surrounding the United States have more consistent and stronger winds that are better for wind-generated electricity than on-shore areas. The problem with offshore wind power is that it is expensive to build 500-foot-tall windmills miles out into the ocean and then string transmission lines through the water. Offshore wind can be two to two and a half times as expensive as onshore wind.[19] As subsidies and mandates for wind power continue, we can expect to see an increase in the amount of offshore wind installation, despite the relatively high cost. The controversies surrounding the Cape Wind project and the Block Island Wind Farm suggest that offshore wind may face social hurdles, in addition to financial ones.

Wind Power's Environmental Benefits May Be Overblown

If climate change and pollution were not a concern, wind power would have very little traction. Many policy makers and members of the public are willing to overlook the economic and reliability issues with wind power based solely on the assumption that wind power is better for the environment than more traditional energy sources. The actual environmental benefits from using wind power are hotly debated, and it can be difficult to separate the hype from reality. One thing that is certain is that wind power does have at least some negative environmental impacts. The debate is over how much better wind is than traditional power sources like coal and natural gas.

Most people assume that wind power has zero carbon or pollution emissions. Wind turbines obviously do not emit pollution or carbon when they are generating electricity, but they do lead to carbon and pollution emissions in more indirect ways. Because of wind power's reliability issues, grid managers have to make sure that cycling reserves are available to compensate for wind's variability. Natural gas, and in some cases, coal, are commonly used as cycling reserves to make sure that energy supply and demand remain in balance. Overall, wind power usually leads to a decrease in carbon emissions, but because of the effects of cycling, wind power is not as green as most people assume.

Scholars and politicians disagree on how significant cycling-caused emissions are. In some cases, depending on what the source of backup power is, cycling does not result in any extra carbon emissions. This can happen when the principal source of cycling is hydropower. In other cases, cycling can partially or even completely negate the reduction in emissions from using wind power. When natural gas or coal is used as the cycling reserves, the environmental benefits of using wind turbines can be severely lessened. The true impacts of cycling have to be evaluated on a case-by-case basis, which depends on how efficient the wind turbines are, how consistently the wind blows in certain locations, and which kind of cycling reserves are available to provide backup power.

Scholars in both North America and Europe have tried to determine the actual environmental implication from using wind power. Most studies find that the amount of avoided fossil fuel emissions from wind power were larger than the added emissions from using fossil fuels as cycling reserves. In 2013, the National Renewable Energy Laboratory concluded that cycling-caused emissions are "negligible" when compared to the carbon reductions of using wind-generated electricity.[20] Many of these studies, however, are based on unrealistic assumptions, which casts some doubt on their accuracy. In some cases, these studies used optimistic assumptions, such as perfect forecasting for wind and electricity demand,[21] high estimates of reliance on clean-burning natural gas turbines for cycling,[22] and optimal usage of transmission capacity.[23] When assumptions are unrealistic, studies can predict results that do not reflect reality.

Using baseload power plants as cycling reserves makes them work less efficiently. Recent studies show that when coal and natural gas power plants are ramped up and down repeatedly, the amount of emissions per unit of generated electricity often increases. Bentek Energy researched the emissions impacts of wind power in parts of Colorado and Texas.[24] When natural gas turbines couldn't provide enough backup for wind power in these two locations, utility providers had to cycle coal power plants. The emission-control equipment for coal power plants operates less effectively when forced to cycle, resulting in more emissions. In some cases, backing up wind-generated electricity can lead to increased nitrogen dioxide and sulfur dioxide emissions. In other cases, increasing the amount of wind power can slightly increase carbon dioxide emissions or leave them effectively the same, depending on the types of electricity generation that are available for backup power. If wind power does not reduce pollution or carbon emissions, there is no purpose in using it. Empirical analysis is necessary to know whether cycling may have more detrimental effects than hypothetical models suggest.

Another environmental impact that many people overlook is the effect of wind turbines on bird and bat populations. Wind turbine blades spin at 138–182 miles per hour, which can easily kill unsuspecting birds and bats.[25] In North America, wind turbines kill an estimated 214,000–368,000 birds annually.[26] In 2012 alone, wind turbines killed an estimated 600,000 bats in the United States.[27] One argument that wind power supporters often employ is that many man-made structures kill hundreds of thousands, if not millions, of birds and bats every year. Cars, skyscrapers, and cell towers all pose deadly hazards. Adding one more hazard that leads to additional wildlife deaths is one of the trade-offs that policy makers and the public should consider when deciding to subsidize and otherwise boost wind power.

Wind power companies have been looking for ways to reduce the mortality rates of these animals. Some companies have been experimenting with radar detection that would slow down or shut off turbines to reduce the number of fatalities. Other companies have tried strike-detection technology, which slows turbines when it detects contact with a bird.[28] Although these technologies may help to save birds and bats from unnecessary deaths, slowing down or shutting off turbines further exacerbates the inconsistency and unpredictability that plagues wind power.

Many people claim that wind turbines are carbon neutral because they do not produce carbon or pollution directly, but that is not necessarily true. The processes of manufacturing, transporting, constructing, and decommissioning a wind turbine itself leads to emissions of carbon and pollution. For a wind turbine to be carbon neutral, it must produce electricity for a minimum period of time until it offsets the amount of carbon that was emitted in its construction. This is called the carbon payback time.

Estimates for carbon payback times vary depending on the calculation method used and the size of the turbine, but most estimates show that the carbon payback time ranges from several months to a few years.[29] Considering that a wind turbine's life ranges from 12 to 20 years, most wind turbines will likely be able to offset the carbon produced in their production, if they remain in operation for their entire estimated life span.[30]

Calculating an accurate payback time is difficult, however, because of the emissions from all the resources that are needed to manufacture a wind turbine. These resources may include electricity, steel, cement, fiberglass, gasoline, diesel, and so on. Wind power comes with unique pros and cons, as with all energy sources. In a market without government interference, utility companies and electricity companies would build as many wind farms as is economically worthwhile. If their customers demanded greener energy, that would boost the demand for wind power without the need for government involvement. But the government is involved in electricity, and supply and demand are only minor considerations when forming electricity policy.

Further, as large amounts of money and time are spent incentivizing wind power, considering the opportunity costs associated with so much public investment in one industry is important. Opportunity costs include all the time, money, and resources that could have been used on other things in place of wind power. It is difficult to say exactly how funding for subsidies would have been spent if they had been allocated to other uses. Policy makers are biased and have limited information, which means they cannot accurately predict what the future of energy will be or should be. Giving special privileges to the wind industry effectively allows it to be the energy winner. Government interference in the market limits competition, which slows the discovery of new and potentially better energy technologies.

CHAPTER FIVE

Solar Power

SunEdison's Taxpayer-Funded Implosion

In 2015, the *MIT Technology Review* declared solar energy company SunEdison as the sixth smartest company of the year, praising the company for "aggressively expanding its renewable energy products and building a business to provide electricity to the developing world."[1] Just one year later, SunEdison filed for Chapter 11 bankruptcy. What went wrong with this supposedly smart company? A deadly combination of government subsidies and bad business practices.

Jigar Shah founded SunEdison in 2003 and strategically formed 20-year power-purchase deals with utility companies across the country. These deals gave investors and electricity buyers more certainty for SunEdison's projects.[2] SunEdison also received $650 million in federal subsidies over the course of the company's short life. In a list of companies that have received the most federal subsidies since 2000, SunEdison came in 13th.[3]

But subsidies are only one part of the story. SunEdison and its subsidiaries pursued almost every avenue possible to extract money from governments at the state and federal levels. In total, various state governments gave SunEdison $30 million in subsidies. New York alone gave the company $21 million. Adding up all the federal loans, loan guarantees, tax-exempt bonds, and insurance, SunEdison raked in an additional $846 million. SunEdison's total support from all government sources added up to $1.5 billion.[4]

In 2014, SunEdison decided to adopt the practice of forming "yieldcos." These are subsidiaries of a parent company that help the parent company lower its capital costs while also funding the parent company's string of projects. Each of SunEdison's yieldcos, called TerraForm Power and TerraForm Global, were supposed to raise cash from public shareholders to finance power projects for SunEdison. The yieldcos would maintain a circular flow of cash to SunEdison, where revenues from SunEdison's power projects would be used to pay debts and shareholders of the yieldcos. Legally, yieldcos are separate companies that can be publicly traded, but in reality, the parent company retains many of the management decisions.

Yieldcos can be a problem if the parent company does not face all the right conditions to make wise decisions. Yieldcos need a substantial cash reserves to make sure that they have enough projects ongoing to generate sufficient revenue. Without strategic foresight, the whole purpose of yieldcos can backfire, and can end up costing the parent company more than if the yieldcos did not exist at all. From 2014 to 2015, SunEdison and its subsidiaries acquired other projects and companies worth $5 billion. These aggressive buy-ups raised the amount SunEdison was paying out in interest expenses and lowered how much cash the company had on hand. SunEdison became more financially strapped due to several lawsuits because the company had promised too much and delivered too little. SunEdison could not keep up with its self-inflicted growth. The company's assets were almost completely drained to pay back growing debt from investors.[5]

An anonymous senior executive at SunEdison described the company's un-businesslike approach to business. He said, "Pipeline took precedence over profitability and high cost leveraged buyouts provided the growth steroid."[6] The company's acquisition frenzy pushed it to buy an ever increasing amount of assets to sustain its growth. Paul Bradley, the chief financial officer of Toronto-based independent power company Northland Power Inc., described SunEdison's style this way: "There was a perceived high-growth rate built in that was not sustainable. It was too good to be true."

The deathblow to SunEdison was its attempted acquisition of Vivint Solar in mid-2015. Vivint would have been SunEdison's most expensive acquisition yet, at $2.2 billion. So far SunEdison had been mainly selling its solar- and wind-generated electricity to utilities, but Vivint's focus was on residential solar panels. Some investors were apprehensive at SunEdison's decisions because many of them did not think homeowners were a stable source of revenue compared to utilities. Other investors wondered if SunEdison had overestimated itself by trying to acquire such a large company. Investors lost faith in SunEdison, and its stock went into freefall.[7]

In early 2016, investors began asking questions about SunEdison's assets and accounting. On February 29, SunEdison announced that it would delay filing its 2015 annual report. Company leaders explained the delay, stating that an internal audit committee was checking on the "accuracy of its anticipated financial position." More lenders backed out of financial support. Vivint also backed out of its agreement to be acquired and instead decided to sue SunEdison for damages.[8]

SunEdison's problems did not stop there. It is now in legal battles, including one from its own subsidiary. SunEdison may have used $231 million to pay off loans, rather than paying for solar developments like it had promised. SunEdison's quest for growth made debt its downfall.[9]

Now the U.S. Justice Department is investigating SunEdison's financing activities regarding its canceled acquisition with Vivint. The Justice Department is also looking into an inquiry from the Securities and Exchange Commission about SunEdison's disclosures to investors. According to the commission, SunEdison may not have been truthful about money it had when its stock prices fell dramatically in 2015.[10]

Since companies succeed and fail all the time, why should we care about the demise of one ill-fated solar company? When their leaders make poor decisions, companies collapse, which is exactly what happened to SunEdison. So what makes SunEdison's case so unique? SunEdison was not just risking its investors' dollars—the company wasted hundreds of millions of taxpayers' dollars by making risky investment after risky investment. Combining all the government-granted subsidies, loans, loan guarantees, tax-exempt bonds, and insurance, taxpayers have been forced to pay hundreds of millions of dollars to a failed company, with no return monetarily or environmentally on their investment. As of April 2016, SunEdison's stock was valued at roughly $176 million, meaning taxpayers paid roughly eight times more than SunEdison was actually worth.[11] When government attempts to pick winners, they often end up picking losers instead. In the process, taxpayers end up losing out too because it is their hard-earned tax dollars that are on the line.

The Basics of Solar Power

The two main types of solar power are photovoltaic and thermal (also known as concentrated solar). Photovoltaic (PV) solar panels are probably more well known. They use materials that absorb light and convert that light into electricity. For the most part, PV panels use silicon crystals to convert light into electricity. Once a panel is exposed to light, it begins generating electricity, which goes through a converter to make the electricity usable. Sometimes a building's solar panels generate more energy than the building needs at a given time, so the excess electricity can be stored in batteries or it can be put into the grid for general use. Batteries are generally uncommon in houses with solar panels because the batteries are expensive and bulky. Until battery technology develops further, batteries for storing solar panel–generated electricity remain a luxury feasible only for the very wealthy.

PV panels are not only for rooftop installations. Throughout the country and across the world, industrial-sized solar plants use PV panels to generate grid-scale electricity. Commercial PV power plants can range from a few acres to thousands of acres. The largest commercial PV power plants in the United States are located in the sunny deserts of the Southwest.

Unlike PV systems, solar thermal power plants generate energy by using the heat of the sun. A large array of mirrors focus the sun's heat onto a thermal receiver, which concentrates the heat and turns water into steam, which drives turbines.[12] Solar thermal systems come in three varieties. Linear concentrator systems are made up of long U-shaped mirrors that focus sunlight onto fluid-filled tubes that run along the interior of each mirror. Once the sunlight heats the fluid in the tubes, the superheated fluid flows to a heat exchanger that boils water into steam to spin the turbine generators.[13]

Another solar thermal system is called the dish/engine, which uses large dish-shaped devices similar to a housetop TV satellite. The mirror dish concentrates light onto a power conversion unit in the middle of the dish. Once the power conversion unit receives the concentrated heat, a fluid-filled thermal receiver powers a small electricity-generating turbine.[14]

The last, and probably most conspicuous solar thermal system, is called a power tower. A large, circular field of flat mirrors focus sunlight onto a tower in the center of the field. The concentrated sunlight heats a thermal receiver on top of a tower, which then transfers the heat to boil water to spin steam-powered turbines.[15]

Although solar power can come in a wide variety of forms and sizes, it has one obvious shortcoming: the sun does not always shine. Nighttime can render solar plants almost completely useless. Winter can significantly reduce the effectiveness of solar plants because the days are shorter in the winter and the sun's rays hit solar panels and mirrors more obliquely. Clouds can also reduce the output of solar plants because not as much sunlight can reach the panels and mirrors to be converted into electricity.

Because the sun does not always shine, the nameplate capacity of a solar plant can be misleading. If a solar plant has a capacity of 100 megawatts, that is the total amount of energy a solar plant could put out at any given instant. On the ideal sunny day, a 100-megawatt solar plant may be able to output 100 megawatts of electricity at a time when sunlight is most direct and there are no clouds. But the same 100 megawatt solar plant will not produce a single megawatt at night.

Solar power, whether PV or thermal, has one of the lowest capacity factors of any major electricity source because it is dependent on factors that are out of any human's control. Capacity factors is a ratio of a plant's actual energy output to its total potential output. The median capacity factor for a utility-scale PV power plant is about 20 percent. For a solar thermal plant, the median capacity factor is about 38 percent. For reference, typical onshore wind power has a median capacity factor of about 37 percent, nuclear plants are about 90 percent, and large coal power plants are about 93 percent.[16] Lower capacity factors indicate that both PV and thermal solar rarely produce their full capacity.

Geography also poses a problem for solar power. Solar energy production is more efficient in areas with a high number of sunny days. California and New York have vastly different climates, which makes solar power much more sensible in California than New York. A person living in Sacramento, California, can expect to see an average of 188 clear days and 3,608 hours of sunlight per year.[17] A resident of Buffalo, New York, would only expect to see an average of 54 clear days and 2,207 hours of sunlight per year.[18] A Sacramento resident with solar panels might produce 6,093 kilowatt-hours a year, but a Buffalo resident with the exact same solar panels would only produce 4,774 kilowatt-hours of electricity a year.[19]

A solar panel's efficiency increases when sunlight hits the panels more directly. Sunlight hits the Earth more directly near the equator and less directly near the poles. The further north a solar panel is located, the less directly sunlight will strike panels. In the winter months, more northerly places in the United States and Europe have much shorter days than nights, making solar panels useless for a large portion of the year. The output of solar panels in colder climates is also limited by snow, which can make solar panels much less efficient when it blocks light from reaching the panels.

For solar panels to work optimally, they need sunlight to strike them directly. Some solar panels have a movable tracking system, which allows them to follow the sun through the day, increasing their efficiency.[20] Although movable tracking systems can help improve the electricity output from solar panels, most residential solar panels are not equipped with these systems. Tracking systems also impose additional costs on commercial solar plants.

Solar Power's Variability Complicates Grid Management

The U.S. electric grid is designed to transmit electricity from power plants whose output can be controlled by power grid operators. Solar power is not easily controllable because its fuel source is at the mercy of the weather and cannot be transported. These characteristics necessitate more complex grid management techniques and new grid infrastructure.

Weather and geography make it difficult for grid managers to plan for solar power accurately. In places that have a large amount of solar power, grid operators try to anticipate how solar power output will change throughout the day. Sometimes grid operators overestimate how much solar-generated electricity will be available, so they have to call on other sources of electricity to ramp up production to keep the amount of electricity in the grid stable. Other times, grid operators underestimate how much solar-generated electricity there will be, so they have to call on other power sources to ramp down their production, and in rare circumstances, grid operators may have to disconnect a solar power plant from the grid to maintain grid stability by reducing power system output.

Being able to accurately predict solar power production throughout the day is important because it has real impacts on both energy producers and consumers. The National Renewable Energy Laboratory (NREL) studied the impact of solar forecasts. Their study estimated that the New England independent system operator could save $46.5 million in generation costs if they could improve solar forecasts by 25 percent.[21] Grid operators and solar power producers can forecast the availability of sunlight by using cloud tracking techniques and numerical weather prediction models. These models use current weather data to predict future weather patterns.[22] For example, the California Independent System Operator (CAISO) uses day-ahead and hour-ahead forecasts for solar power to make sure that the electricity supply consistently meets demand. But even hourly predictions are not enough to fully compensate for solar power variability, so CAISO is working to forecast solar output at 5-minute intervals.[23]

Similar to wind power, connecting solar plants in different locations is one way to combat the problem of variability. Because weather is more predictable over larger areas, grid operators will sometimes connect several solar plants to increase the predictability and consistency of electricity generation. Interconnection of solar plants can reduce the need for backup power if a specific location's solar plants do not produce enough to meet demand because other solar plants may be able to make up the difference. The Department of Energy has argued that aggregating solar will decrease fluctuations in solar power output, but insufficient data does not allow us to accurately quantify the actual effect of connecting solar plants, or what the cost of doing so would be.[24]

When solar power fails to produce enough energy, and interconnected solar plants are not an option, operating reserves become necessary to keep the amount of electricity stable.[25] Most large-scale power plants, like coal and nuclear, are not designed to ramp up and down quickly for variable renewable energy sources like wind and solar power. When backup power sources need to be cycled to compensate for the variability of wind and solar power, the power plants providing backup power have higher operation and maintenance costs.[26] When power providers incur greater costs, they have no choice but to pass those costs along to their customers.

One unique issue with solar power is a problem with over generation and under generation in a single day. In the middle of the day, solar power plants produce the most energy, but electricity demand is usually the lowest during the times when solar plants are most productive. In places with a high concentration of solar power like California, baseload power plants have to decrease their production to maintain balance on the grid.[27,28] Ramping down baseload power plants becomes more common in areas where solar power makes up a bigger portion of the electricity supply.[29]

In the evening, solar power plants produce less electricity as the sun starts to go down. Unfortunately, demand for electricity peaks in the evening for most places. Traditional power plants, like coal and nuclear power, must compensate for the decline in solar power production as the sun goes down. Unfortunately, these large plants usually need several hours to ramp up sufficiently to fully meet the peak electricity demand in the evening.[30] As solar power grows in many areas, grid managers will have to be more concerned with the problems of over generation and increased ramping. For example, CAISO conducted a study that predicted how increasing solar power production will affect ramping. CAISO estimated if solar power continues to grow at the same pace until 2020, grid operators will have to shift 13,000 megawatts of electricity from solar power in the day to other power sources in the evening.[31] Ramping up 13,000 extra megawatts of electricity within roughly three hours is a nightmare for grid operators who are used to managing power plants that provide a fairly constant baseload of power throughout the day.

The variability of solar power could be solved by grid-scale energy storage, but such large-scale storage is not economically feasible. Although there are many types of energy storage that could alleviate the variability of intermittent sources of electricity, like wind and solar, most of these solutions cost too much to implement across an entire power grid. For short-term storage possibilities, the most obvious choice for grid managers is traditional batteries. Batteries, however, are expensive and have environmental costs when they are manufactured and disposed of. Traditional batteries are not used for grid-wide energy storage anywhere in the United States simply because they cost too much and take up too much space.

Other, more innovative solutions to short-term energy storage include flywheels and high-energy capacitors. Flywheels are large disks that use excess electricity from solar power to make them spin, and when the sun is not available for solar power, the momentum in the spinning flywheels can be used to spin electricity-generating turbines. Another promising technology is high-energy capacitors, which stores energy as an electrostatic field between two conducting plates that can release electricity when there is not enough sunlight for solar power.[32]

For more long-term storage, energy innovators are looking to pumped hydroelectric storage and compressed air energy storage to solve the inconsistency problems with solar power (as well as wind).[33] Pumped hydroelectric storage uses excess solar-generated electricity to pump water into a reservoir or a storage tank. When the sun is not shining, the stored water can be released through a hydroelectric turbine to generate electricity. Compressed air energy storage uses excess electricity from wind or solar power to pump air into pipes or caverns underground. When solar power is not available, energy producers can release the compressed air to spin small-scale wind turbines for electricity generation.

Although these solutions may help compensate for the intermittency issues with solar power, none of them have been implemented on a large scale, since the growth of the solar industry is outpacing the innovators who are working to solve its problems. As the government continues to subsidize and mandate solar energy, its intermittency issues will continue to grow, and a grid-wide innovation will be necessary before intermittent renewable energy sources begin impacting the reliability of the electric grid overall.

Further complicating energy storage, whether large or small scale, is that utility companies hesitate to invest in storage in the face of government regulations. Many utility companies choose not to invest in storage until they are certain which regulations will affect energy storage projects and how those regulations will affect profits.[34] For example, in 2013 the California Public Utilities Commission set a mandate to have 1.3 gigawatts of energy storage capacity available for the state's grid by 2020.[35] Southern California Edison, one of the state's utility companies, began working toward this mandate by investing $50 million. This investment was meant to provide 8 megawatts of power storage with lithium-ion batteries.[36] But Southern California Edison soon learned that energy storage mandates are much easier to reach in theory than in fact. After spending $50 million to develop just 8 megawatts of storage capacity, the utility company warned that California's energy storage mandate "could cost up to $3 billion with uncertain net benefits for customers."[37]

Federal and State Subsidies Boost the Solar Industry

For decades, governments at the federal, state, and local levels have mandated and subsidized solar power to boost the industry, even though solar power is intermittent and expensive. Even with billions of dollars, decades of time, and legal mandates for its use, all forms of solar power still generated less than 1 percent of the electricity in the United States in 2013.[38] One of the main reasons that the solar power industry has become so dependent on government assistance is because it is one of the most expensive sources of energy relative to its output.

Calculations for the levelized cost of electricity (LCOE) try to measure an energy source's total average costs over its lifetime. These costs include construction, fuel, operations, maintenance, and efficiency. Although LCOE calculations are often biased or miscalculated, nearly every LCOE calculation shows that solar power is more expensive than most other forms of electricity generation. At first glance, solar power may seem economical because it does not have any fuel costs, but when considering the lifetime costs, the truth is much bleaker. The Energy Information Administration found that PV solar power's levelized cost is $125.3 per megawatt-hour, and solar thermal power's levelized cost is $239.7 per megawatt-hour. For reference, the levelized cost for a newly constructed coal plant is approximately $95.1 per megawatt-hour, while a natural gas combined cycle plant costs roughly $75.2 per megawatt-hour.[39]

The federal government has not mandated the use of solar power, but many states have renewable portfolio standards (RPS) that legally require a state to produce or consume a certain amount of electricity from renewable sources. For states that have enacted RPS requirements, some have specific mandates for solar power. For example, 20 states have specific solar mandates.[40] Some states' lawmakers were more ambitious than others. New Jersey's RPS mandates that 4.1 percent of the state's electricity must come from solar power by 2028.[41] North Carolina, on the other hand, only requires 0.2 percent solar generation by 2018.[42]

Although people with good intentions may support government mandates for renewable energy, these mandates often impose higher costs on taxpayers and electricity consumers. Not only do they raise costs, but most states have a difficult time even achieving their RPS mandates for many years. For example, California's RPS required that the state must reach 20 percent of its electricity from renewables by 2010. At the end of 2009, however, California only produced 13 percent of its electricity from renewable sources.[43] Similarly, New Jersey's RPS law required that 6.5 percent of the state's electricity must come from renewables by 2009. By 2009, New Jersey had only reached 1.53 percent renewable generation.[44]

Even when states have enough tax dollars to fund their solar programs, the benefits may exceed the costs. On the surface, solar mandates and subsidies are put into effect for environmental reasons, but because solar power is such an expensive energy source, the amount of public money spent on solar power is far more than the savings from pollution and carbon emissions. A study from the George Washington Institute of Public Policy analyzed the effects of solar incentive programs in 10 states. The study found that the solar incentive programs in effect in these 10 states would only reduce carbon dioxide emissions by 6.1 million metric tons over a 20-year period.[45] To put this into perspective, 6.1 million metric tons of carbon dioxide is less than the emissions from two average coal power plants in a single year.[46] These 10 states are putting billions in tax dollars into a system that does very little to actually affect carbon emissions at all. The opportunity cost of using public money so inefficiently has real implications for the well-being of citizens and the environment.

In Louisiana, legislators passed a solar energy tax credit in 2008 that covers up to 50 percent of the first $25,000 in installation costs for residents who install solar systems.[47] Before it was fixed, a loophole in the law used to allow state residents to install multiple arrays on their homes and to accept a separate tax credit for each one.[48] Louisiana's solar tax credit has resulted in $151 million in lost revenue since 2008.[49] With several years of data now available, Louisiana's program has cost 122 times more than the highest estimates when the program was first proposed in 2007.[50]

Talking about the realities of solar power, or any renewable energy source, is often controversial because people do not always understand the implications and trade-offs involved. Academics and researchers disagree on how much the government should be involved in the energy sector. They also disagree about the effectiveness of government policies at doing what they were intended to do. When it comes to solar power, several studies show different results in how effectively state-level incentives have increased solar power production. Because California's solar market is one of the largest in the United States, several universities have studied the effects of California's initiatives to boost the solar market. Researchers at the University of Colorado Boulder found that the California Solar Initiative (CSI) and its associated rebates were directly responsible for 58 percent of all residential and commercial solar panel installations in California between 2007 and 2012.[51] Researchers from Vanderbilt University and Sandia National Laboratories, however, came to a nearly opposite conclusion. These researchers compared California's solar market to a similar market without solar incentives, and found that the program had only a small impact on solar adoption rates.[52]

Although the effectiveness of the CSI is debated, the initiative has likely increased solar installments to at least some degree. Hundreds of millions of dollars have been spent on the CSI, but the emissions reductions from the program have been miniscule compared to the costs that the CSI has imposed on California's taxpayers. The University of Colorado Boulder study found that the CSI did little to reduce carbon or pollution emissions. The study estimated that CSI only led to a reduction in carbon dioxide and nitric oxide by roughly the same amount that a small to midsize natural gas power plant would have produced.[53] Natural gas power plants are the cleanest fossil fuel power source, so California taxpayers have been forced to spend $437 million on a program that eliminates a relatively small and inconsequential amount of emissions. Policy makers and the general public in California will have to begin asking themselves if the millions of taxpayer dollars going to solar power subsidies actually provide them with the results they want.

The federal government also subsidizes solar power through several multi-billion-dollar programs. In 2012, officials in the Government Accountability Office (GAO) investigated whether different federal agencies had overlapping renewable energy initiatives. After collecting data from 24 federal agencies, the GAO found that these federal agencies employed almost 700 renewable energy-related initiatives. Of these nearly 700 programs, 345 of them included at least some provision for solar power.[54,55]

One of the most influential federal programs for solar power has been the solar Investment Tax Credit (ITC). Enacted in 2006, the solar ITC reduces federal income taxes for owners of qualified solar power systems. If a person installs a solar power system before December 31, 2016, the ITC allows 30 percent of expenditures on the system to be claimed against federal taxes.[56] The solar ITC applies to both residential and commercial solar projects.[57] When calculating the total value of the ITC tax credit, residential or commercial solar developers can include equipment and labor, but they cannot include structural components, like platforms for the solar equipment.[58] Other renewable energy projects also have an ITC, including small wind turbines, fuel cells, and geothermal systems.[59]

The program has proven successful at boosting the solar industry, and since the program began, solar installations have increased at an average of 76 percent per year.[60] Although this tax credit has helped the solar industry grow, this growth has been dependent on government subsidies. If Congress had not extended the ITC before its initial expiration at the end of 2016, the Solar Energy Industries Association calculated that the expansion of installed solar capacity would have gone down by 57 percent.[61]

Throughout 2015, the expiration date of December 31, 2016, loomed over solar power developers who wanted to take advantage of the free money guaranteed by the ITC. The uncertainty about whether the ITC would be renewed made solar power developers become skittish. If they were not able to complete their projects in time to qualify for the government handout, many projects may no longer be worth it. When many developers saw that they would potentially not receive ITC money before the expiration, they completely canceled their projects, rather than face the economic burden of solar power on their own.[62]

The solar industry's dependence on the ITC became clear in the two years before the ITC was initially set to expire at the end of 2016. For example, BrightSource Energy, Inc., canceled a utility-size solar project in 2014 because company leaders were uncertain if their project would be done in time to receive ITC money. Without the financial incentives of the ITC in place, solar power projects become much less feasible because developers must make investments with their own money. Their profits are no longer guaranteed by American taxpayers. To avoid losing out on the economic advantages of the ITC, solar industry advocates and other renewable energy industry leaders lobbied Congress to extend the ITC.[63] By December 2015, the lobbying efforts paid off. In the Consolidated Appropriations Act of 2016, Congress renewed the ITC for any project that comes online by the end of 2023.[64]

The ITC bolsters the solar energy industry and makes solar power artificially cheap at the expense of American taxpayers. The GAO estimates that the ITC has cost taxpayers $4 billion in lost revenue between 2000 and 2013.[65] The GW Solar Institute at George Washington University found that solar power would be significantly more expensive without taxpayer funding. They found that if solar power could not benefit from the ITC and the depreciation advantages in the tax code, the price of a standard 20-megawatt PV solar plant would be 58 percent higher.[66]

Another federal program that helps the solar industry is the Modified Accelerated Cost Recovery System (MACRS). This program, enacted in 1986, grants owners of renewable energy systems, including solar plants, a five-year recovery period for depreciation costs related to renewable energy equipment.[67] MACRS also allows companies to recover some of the capital costs of a solar energy facility. In essence, MACRS speeds up the rate of return for solar investment, meaning that they become profitable more quickly.

The US Partnership for Renewable Energy Finance (US PREF) found that removing the MACRS program would make the solar industry much less profitable. The US PREF estimates that a solar power project earning a 7 percent rate of return under MACRS would only earn about a 5.3 percent return on investment without the program.[68] The success of the solar industry under MACRS also comes at the expense of taxpayers. According to the GAO, MACRS-related revenue losses amounted to $1.7 billion from 2000 to 2013.[69]

On top of subsidies, tax credits, and an accelerated depreciation program, the federal government has also offered loan guarantees to solar power companies. Section 1705 of the American Recovery and Reinvestment Act of 2009 (ARRA) authorized loan guarantees to qualifying renewable energy projects that began construction before September 30, 2011.[70] When solar power companies cannot repay lenders for loans, the loan guarantee program allows the DOE to repay lenders if the borrower defaults. This means that if solar companies default on their loans, they do not bear the costs. Instead, taxpayers foot the bill. The loan guarantee program has become one of the most controversial federal policies for renewable energy sources. When factoring in interest payments on loans, the amount paid out for defaults, and the borrowing costs owed to the U.S. Treasury, the DOE loan guarantee program has cost taxpayers hundreds of millions of dollars.[71]

Another federal policy that boosts solar power development was Section 1603 of the ARRA, which set up a Treasury grant program to increase renewable energy development.[72] Under this program, energy developers could choose a direct cash payment from the Treasury instead of a tax credit. The 1603 program could be used in place of the ITC or the production tax credit (PTC). The 1603 program was in effect from 2009 to 2013, and the Treasury handed out almost $20 billion during that time. Of that $20 billion, $4 billion went to solar projects specifically.[73] Companies that began construction on eligible projects before the expiration date can still receive grants from the 1603 program. As of January 2015, the 1603 program had granted a total of $7.8 billion for solar projects.[74] These payments represent transfers of wealth from American taxpayers to favored solar power companies.

Net Metering Policies

States have also enacted policies to encourage the development of solar power. One of the most controversial of these policies is net metering, which allows owners of residential solar panels to get paid for extra power they send back into the electric grid. As of 2016, 44 states, plus the District of Columbia and four U.S. territories, have passed net metering policies. Under net metering laws, utility companies are required to purchase electricity from residential solar panels.[75] Net metering customers usually receive the retail price for any electricity they sell back to the grid.[76,77] In essence, net metering customers have their electricity meters spin backward when their solar panels produce more electricity than the house consumes. If a net metering customer generates more electricity than they consume over a month, they could actually make money off their solar panels.

Like other policies intended to encourage production of electricity from renewables, net metering results in unintended consequences that transfer wealth and disproportionately harm low-income Americans. Net metering financially benefits those who can afford to install solar panels on their houses, but also has the unintended consequence of raising electricity costs for people without solar panels. The grid was designed for a one-way flow of electricity. When net metering occurs, the two-way flow causes mechanical stress on the electric grid. Many states do not allow utility companies to charge residential solar panel owners for the added maintenance that is necessary under net metering policies. When net metering customers are paid the retail rate, they are not paying the fixed cost of using the grid in the first place. Their neighbors have to subsidize those fixed costs through higher electricity rates.

To account for the lower prices paid by net metering customers, utility companies raise electricity prices for everyone. Net metering policies are regressive because people without solar panels end up subsidizing their wealthier neighbors with solar panels through higher electricity rates.[78] In addition to being regressive, net metering may not be as cost-effective as most people hope. The California Public Utilities Commission estimates that net metering in the state will cost non-solar customers $1.1 billion by 2020.[79]

Another state policy intended to encourage renewable energy is known as Feed-in Tariffs (FIT). FIT programs begin with a contract between residential solar panel owners and utility providers. In this contract, the utility provider sets a fixed price for the electricity that solar panel owners generate and put back into the grid. These contracts are usually good for 15 to 20 years.[80] The money to pay for FIT programs often comes from higher utility bills of electricity consumers who do not have solar panels. Once again, these policies end up being regressive as all utility customers end up subsidizing residential rooftop customers.

In 2009, the city of Gainesville, Florida, passed the first FIT program in the United States. Gainesville's program instituted 20-year contracts, and for the first year of each contract, Gainesville Regional Utilities (GRU) drafted contracts to pay residential solar panel owners 32 cents per kilowatt-hour.[81] For comparison, the average electric rate in Gainesville in 2016 was just under 13 cents per kilowatt-hour.[82] Each year after, the contracts stated that GRU would gradually decrease the rate for payments for solar-generated electricity.[83] Since the program began in 2009, the city's solar capacity grew by 18 megawatts, which is relatively small considering that a single, average-sized coal power plant has a capacity of roughly 500 megawatts. Although Gainesville's program achieved its goal of spurring additional solar power capacity in the area, the program also led to some of the highest electricity rates in the state.[84] By December 2013, Gainesville's FIT program participants had received a total of $11.4 million from their non-solar neighbors.[85] Gainesville's electricity rates became the most expensive in Florida. To quell discontent regarding the high rates, Gainesville's city commissioners chose to suspend the FIT program in December 2014.[86] The 20-year contracts, however, are still valid, and GRU must continue to pay people who had already signed onto the FIT program. That means non-solar customers will continue to subsidize their neighbors for years to come. Estimates show that Gainesville ratepayers will be liable for a total of $74 to $84 million over the next 20 years until the FIT contracts expire.[87]

Some utility companies have realized the shortcomings to their net metering and FIT programs and have tried to enact reforms. Arizona's Salt River Project (SRP) reformed its net metering policies in 2015 after many of their non-solar customers complained that they were footing the bill for additional operation and maintenance costs caused by net metering. Before the reforms, SRP calculated that its customers without solar panels collectively paid an extra $9 to $10 million each year to make up for the costs of net metering. Even after shifting costs onto non-solar customers, the program faced an estimated net loss of $46 million. To make sure solar customers are paying for their use of the electrical grid, the Salt River Project increased rates for net metering customers by adding a $50 surcharge each month. Non-solar customers will also have to pay approximately $5 on top of their monthly bill. Although rate increases for solar customers were controversial, the Salt River Project obtained the support of low-income advocates within the state as well as economic advisers for its attempts to shift some of the costs of using the electrical grid back onto solar customers.[88] Determining the correct rate that should be charged for net metering customers is difficult. The one thing that is clear is that net metering policies, like so many other well-intentioned state-level energy policies, have unintended consequences.

The Environmental Impacts of Solar Power

Although policies enacted to boost solar power vary in their implementation, almost all of them are justified by solar power's supposed environmental benefits. The underlying policy goal is to reduce production of electricity from traditional sources of energy in favor of renewables, thus creating a net environmental benefit. If traditional energy sources did not have environmental impacts, then there would be no real push for solar power. Although solar power has some environmental benefits, in many cases it is not as environmentally friendly as many people believe it to be. In many cases, the environmental implications are hidden, and solar power is less beneficial than most people assume.

One of the most important environmental implications of solar power is water consumption. All solar plants, whether PV or thermal, need water for cooling or cleaning. A utility-size PV system requires roughly 7 million gallons of water each year for maintenance, including washing dust off of panels.[89,90] Over the lifetime of a typical PV plant, it can consume about 520 gallons of water per megawatt-hour.[91] Solar power consumes less water than coal, natural gas, and nuclear plants per unit of energy generated, but solar facilities are often built in water-scarce areas, which can lead to problems.[92]

Because thermal solar plants use steam turbines, they usually require more water than traditional PV solar panels, which only require water to be cleaned. Solar thermal plants can use wet-cooling or dry-cooling technology to generate electricity, but both technologies come with their own pros and cons. Wet-cooling technology is the most common, and it uses roughly 1,240 gallons of water per megawatt-hour of energy produced.[93] Coal, nuclear, and natural gas plants also use wet-cooling systems to generate electricity, but solar thermal power plants with wet-circulation cooling systems use more water for every unit of electricity compared to traditional energy sources using the same technology.[94]

Dry-cooling systems are becoming more popular due to concerns over water consumption, but dry-cooling systems are less efficient than wet-cooling systems. Solar power developers in Idaho, Arizona, California, and Nevada are now using dry-cooling systems to avoid depleting already scarce water resources.[95] As the name implies, wet-cooling systems use water to cool the steam from turbines, but dry-cooling systems use the air outside the power plant to condense the steam. Although it uses 90 percent less water than a wet-cooling system, dry-cooling systems can result in a 7 percent drop in electricity production and a 10 percent increase in capital costs.[96]

Dry cooling does not work as well when the surrounding air temperature is consistently above 100 degrees Fahrenheit because the air is too warm to condense the steam back into water.[97] In the hot and arid Southwest, dry-cooling thermal power plants run even less efficiently than in other regions.

Another environmental concern is emissions. Because solar power is intermittent, grid managers have to cycle non-solar power plants to make up for when solar does not produce enough to meet demand. This extra cycling also results in extra emissions. Depending on the source of backup power, cycling conventional power plants can increase emissions per unit of electricity.[98] Although cycling is an important concern, the National Renewable Energy Laboratory has concluded that cycling-caused emissions are still less than the amount of carbon that is reduced by using solar power.[99]

Cycling emissions, however, are not the only indirect source of emissions to consider. Carbon and pollution emissions also occur from the process of manufacturing, transporting, installing, maintaining, and decommissioning solar facilities.[100] The amount of carbon and pollution created during these stages is difficult to measure because each circumstance is unique. These indirect emissions, however, cannot be ignored if the policy goal is to reduce environmental impact.

The manufacturing process for solar panels has also been controversial because of its impacts on environmental quality and human health. In 2012, China manufactured over 60 percent of all PV panels in the world.[101] Chinese-made panels have a carbon footprint twice as big as those made in Europe.[102] Toxic chemicals, such as silicon tetrachloride, hydrochloric acid, and hydrofluoric acid, are needed to produce solar panels, and PV factories in China often dump these chemical in rivers and fields near their facilities.[103] Multiple news outlets have covered stories of solar power manufacturers in China polluting waterways and air, leading to the death of fish and livestock, as well as inflamed eyes and throats for people living near the factories.[104]

For example, Luoyang Zhonggui High-Technology Co., located in the Henan Province, produces polysilicon for solar panels that are exported to many countries across the world. The *Washington Post* found that Luoyang Zhonggui had been dumping toxic materials near residential and agricultural areas. An independent, nationally accredited laboratory analyzed soil from dump sites near the factory, and the tests revealed the presence of chlorine and hydrochloric acid. These substances do not occur naturally in the local soil, but they do come from silicon tetrachloride, one of the factory's main waste products. Li Xiaoping, deputy director of the Shanghai Academy of Environmental Sciences, told the *Washington Post*, "Crops cannot grow on this, and it is not suitable for people to live nearby." Some nearby residents, including local businessman Zhang Zhenguo, theorize that the government is turning a blind eye to the pollution. He told the *Washington Post*, "They didn't test [the soil] because the government supports the plant."[105] China exports approximately 95 percent of its panels to other countries, but the Chinese people are left to deal with these environmental problems.[106]

Further environmental concerns are the effect of solar power plants on wildlife. For example, the Topaz Solar Farm in California was built in the habitat several federally protected species, including the San Joaquin kit fox, the golden eagle, and the giant kangaroo rat. Topaz worked with the United States Fish and Wildlife Service on a conservation plan to minimize the impacts of the solar plant on wildlife populations.[107] Even with mitigation techniques, there will still be effects on the wildlife in the area. Trade-offs are inevitable, and in Topaz's case, wildlife was the trade-off.

Wildlife also has an effect on the economics of solar farms. The Ivanpah Solar Electric Generating System, located in the California desert near the Nevada border, has displaced 144 desert tortoises. The desert tortoise is on the federal government's threatened species list. It is also the designated state reptile of Nevada and California.[108] Ivanpah's developer, BrightSource Energy, has spent over $56 million capturing and relocating these tortoises over the solar plant's life.[109] These costs, as with any costs, are passed on to electricity consumers.

As with all energy sources, solar power comes with trade-offs. Policies enacted to boost production of electricity from renewables have good intentions. But unfortunately, policy makers do not have perfect information about what the impacts of their policies will be. Solar power companies are incentivized to lobby the government for preferential treatment because they stand to gain financially. When these problems interact, they often cause well-intentioned policies to have negative and unforeseen consequences. In the case of solar power, these consequences involve higher electricity rates for some, lower reliability in certain cases, and a higher tax bill for all Americans.

CHAPTER SIX

Biomass

British Policies Burn American Trees

Burning trees to help climate change is on its face an oxymoron. Biomass power plants around the world, however, claim to do just that. One of the strangest examples of the global biomass industry began in England in the early 2000s. Drax Power Station, England's largest power plant, has burned coal since the 1970s, but in the past few years the power plant has switched to burning mainly biomass. When most people think of biomass, they do not think of burning trees. And when most people think of burning trees, they do not get warm fuzzy feelings about combating climate change. Drax Power Station, however, is doing just that, by burning trees imported from the United States to generate "renewable" energy.

In 2006, when Drax still burned primarily coal, it became a target for climate change activists. Roughly 600 people showed up to what was hyped as "The Battle of Drax." The so-called battle turned out to be more of a strange festival intermingled with a few arrests. One of the protesters was even dressed as a clown, who ran around yelling, "Sustainable energy is fun—really fun—I promise it is." The majority of the climate change protesters were camped in a farmer's field within view of Drax's looming chimneys and cooling towers. Local police forces had stationed themselves at the power plant in case of illegal trespass. A group of the demonstrators forced themselves through a line of police officers who were blocking the small lane that leads to the power station. Police ended up arresting 38 people, most of them for trespassing.[1]

The controversy over Drax's coal burning did not end with the 2006 protest. In 2008, a group of roughly 30 people from the Camp for Climate

Action stopped a train carrying coal to Drax and shoveled the contents of the train onto the tracks as a protest against the evils of climate change. The activists stopped the train by dressing up as rail workers in fluorescent jackets waving red flags that signal train operators to stop. The train stopped on an iron girder bridge where the activists had stashed food, water, bedding, and even a mobile toilet. One activist said that they had enough supplies to keep up the protest for at least two days. The activists climbed aboard the freight train and began to shovel over 20 metric tons of coal onto the tracks. The group also draped a banner over the train, saying "Leave it in the ground."[2]

Police arrived soon after the train was stopped, and a day-long standoff ensued. Police officers, many dressed in riot gear, surrounded the train. The protesters were well prepared, however. Many of them had switched out the Sim cards in their cell phones to avoid interception. The coal shovelers had masks to protect against coal dust. Other protesters chained themselves to the train. One protester had even dressed as a canary and carried a sign that said "How many warnings do we need?"[3,4]

The police ended up arresting and charging 22 of the train protesters with violating the Malicious Damage Act of 1861. Nearly a year later, the 22 people who had been charged used their trial as another protest stage. The defense brought in university lecturers, filmmakers, and others to speak to the jury about global warming and Arctic melt and the history of reforming Britain's laws. The judge had asked the defendants over 20 times to return to the subject of the train, but Dr. Paul Chatterton, a senior lecturer in geography at Leeds University, and the defendants ignored the instructions and continued to explain the "deadly and urgent threat" of Drax's carbon emissions.[5,6]

Despite the defendants' heartfelt arguments about the dangers of climate change, the jury ultimately decided that they had illegally stopped the coal train. Five of the defendants who had previous convictions were sentenced to 60 hours of community service. Three others were sentenced to pay £1,500 in costs and compensations to Network Rail, the owner of the train. The other members of the group received conditional discharges for 12 months. Judge James Spencer explained to the defendants, "You were involved in an elaborate plan to interfere with other people going about their lawful business. Each one of you were involved in this scheme to disrupt the influx of coal to Drax power station."[7]

In 2013, the combination of bad publicity surrounding coal combined with government subsidies for renewables led Drax officials to switch from coal to biomass. Drax's Chief Executive Officer Dorothy Thompson said, "We see a key part of our future as converting from essentially a coal station to a biomass station. It will take Drax from being the largest carbon emitter by site in the U.K. to being, probably, one of the largest renewable plants in the world." Thompson failed to note, however, that burning biomass emits more carbon dioxide than coal.

Two problems have arisen since Drax has switched much of its electricity production from coal to biomass. First is the environmental impact of burning biomass. Second is the economic consequence of being dependent on government subsidies for renewable energy.

In 2013, Drax burned 7 million metric tons of plant material. About 90 percent of that material was timber shipped several thousand miles across the ocean from the United States on diesel-powered boats.[8] Unfortunately, burning biomass can produce one and a half times as much carbon dioxide than burning coal. In many cases, biomass also emits more ash, nitrogen, and other pollutants than coal.[9]

Members of Parliament became concerned by how much biomass was being imported from the United States, and in late 2014, they began to consider removing biomass from renewable energy incentives. With the prospect of less profit from government subsidies, investors became skittish, and Drax shares dropped sharply.[10,11] Only a month after the U.K. planned to stop subsidizing Drax, the British government reversed its position and offered Drax a contract guaranteeing payment for every unit of renewable electricity produced.[12] With the certainty of government money, Drax's shares rose almost 6 percent.[13] In 2015, Drax received £4.3 billion ($6.7 billion) in subsidies to burn almost 8 million tons of biomass material.[14]

When Drax was burning coal, protesters fought against the harms of climate change. Unfortunately, subsidizing biomass has likely led to even more emissions from the Drax Power Station, despite the public perception of biomass as renewable. Without government support, Drax's biomass endeavors would be much less profitable. Government officials are enabling this power plant to profit and encouraging even greater environmental damage than would have happened if Drax had just continued to burn coal. In the future, activists may hijack trains filled with biomass to protest this new source of carbon emissions.

State and Federal Policies for Biomass

Politics often gets in the way of good ideas. Biomass policy in the United States is just one example of how the political process has transformed a reasonably good idea into a confusing, illogical, and complex policy. Biomass is any organic material, and we can use biomass as a fuel to generate electricity, much like coal. Biomass is one of the most diverse and unique electricity sources. Energy producers take materials like wood, grass, or garbage, and burn them to create steam. The steam then spins turbines, which generates electricity.

Before the government got involved in the biomass industry, energy producers would use biomass on a small scale to generate electricity. Waste products, like wood scraps, crop residues, and garbage, did not have any other economic value. Energy producers would use this waste, which was not useful for anything else, and burned it to make electricity. In most cases, this waste was burned to supplement coal-fired power plants.

Since the beginning of the environmental movement in the 1960s, climate change and environmental quality have taken center stage in American politics. Beginning in the early 1990s, Congress began to enact renewable energy policies that were meant to give renewable energy industries a boost. Wind and solar power have become the most prominent and recognizable renewable energy sources, but biomass also received special treatment under federal law. Biomass became eligible for federal subsidies, and so the market for biomass began to grow. Throughout the 1990s and into the 2000s, many states began adopting renewable portfolio standards (RPS), which mandated that a minimum percentage of a state's electricity must come from renewables by a target date in the future. Each state's individual RPS mandates are unique, but the mandates also helped speed up the growth of the biomass industry, as well as other renewable energy industries.

The RPS requirements in some states mandate that a particular amount of electricity must come from a specific energy source. These source-specific mandates are called carve-outs. Many states have carve-outs for wind or solar power, which are the most politically popular and most subsidized forms of renewable energy. A few states, however, have carve-outs for biomass. For instance, Minnesota's RPS requires 110 megawatts of biomass power.[15] Even without specific carve-outs, biomass is still eligible to satisfy general requirements for renewable energy in all states with RPS requirements.[16]

Although most RPS requirements are 10 to 20 years old, many states are having difficulty reaching the mandated levels of renewable energy, especially for biomass. In 2007, North Carolina passed a new carve-out for biomass that required 900,000 megawatt-hours of electricity come from poultry waste by 2014 and 0.2 percent of the state's electricity come from swine waste by 2018. Soon after the new carve-outs were adopted, electricity producers in North Carolina realized these new standards were too strict.[17] In 2015, North Carolina changed its RPS requirements to delay the biomass carve-outs by two years. North Carolina energy producers must now generate 900,000 megawatt-hours of electricity from poultry waste by 2016 and 0.2 percent of the state's electricity from swine waste by 2020.[18] The North Carolina Utilities Commission extended the requirements because electricity producers could not realistically generate the mandated amounts of electricity from swine and poultry waste.

The high capital cost of methane-to-electricity systems also made the RPS carve-out for biomass a huge economic burden to electricity producers. Not only were the costs a problem, but utility companies were not even sure which environmental regulations applied to their facilities, making them even more hesitant to move ahead with waste-based electricity projects. The North Carolina Utilities Commission that helped to modify the state's biomass carve-outs reported that "[the] costs [of installing biomass projects] are eventually borne by retail consumers, and allowing more time for the market to develop is in the public interest in that it will protect customers from the inflated cost of an undeveloped industry."[19] Despite the fact that the commission admits that biomass mandates increase electricity prices for average citizens, biomass carve-outs are still on the books.

In theory, subsidies are intended to help a technology become competitive against other technologies. From 2009 to 2013 alone, woody biomass received $650 million in subsidies.[20] After years of government support, the biomass industry still demands an ever increasing amount of subsidies.[21] Despite its intentions, subsidies do not always make one industry more competitive in the market. The lure of subsidies entices companies and becomes a source of profit that no one would ever willingly turn down.

The production tax credit (PTC) is the largest federal subsidy for renewable energy, including most types of biomass.[22] Not all biomass receives the same amount of money under the PTC. Closed-loop biomass, which comes from energy crops, receives a tax credit of 2.3 cents per kilowatt-hour. Open-loop biomass, which comes from farm waste, forest waste, and landfill gas, receives 1.1 cents per kilowatt-hour.[23] Even before subsidies were put into place, some electricity companies used open-loop biomass to generate electricity when it was economically prudent.[24] The PTC and other subsidies allow biomass companies to gain more money at the taxpayers' expense, even though many companies were already profitably using biomass without the government's help.

In addition to the PTC, the Biomass Crop Assistance Program (BCAP) subsidizes farmers to harvest biomass waste that can be used to generate electricity. The U.S. Department of Agriculture uses the BCAP to match payments that biomass energy companies make to farmers or other biomass waste producers. The BCAP matches payments up to $12.5 million per year.[25] The U.S. Forest Service claims that the BCAP helps to generate energy, restore watersheds, and remove dead wood that could fuel forest fires. In 2014, the BCAP helped to remove 200,000 tons of biomass from federal lands.[26]

The BCAP has trade-offs for environmental health. Burning forest wastes for electricity contributes to carbon emissions and air pollution, but limiting the size and scope of wildfires may actually cause a net decrease in pollution and carbon emissions. It is difficult, if not impossible, to accurately quantify the amount of emissions from electricity generation to avoided emissions from wildfires, but thinking about trade-offs is an important factor in making beneficial energy policy.

Environmental effects aside, the BCAP has become a financial crutch to many companies in the biomass industry that could not exist without government funds to prop them up. According to the president of the Biomass Power Association, "seven of Maine's ten biomass energy plants would have shut down without the new influx of funds."[27] Without the BCAP, many people would not choose to enter the biomass industry. Without government support, the biomass industry could not maintain its current size if it lost government support.[28]

With the combination of subsidies and mandates, the biomass industry has expanded rapidly for the past two to three decades. Thanks to the government's help, most biomass energy producers no longer use only waste products. Government policies promoting the use of biomass have made nonwaste biomass resources, including lumber, economically viable as fuel to generate electricity.

Although renewables are often lumped into one category, they each function very differently. Biomass is one of the most reliable forms of renewable energy because it can be burned to produce electricity whenever it is needed. Biomass, unlike wind and solar power, can produce baseload power because it is not intermittent. On a small scale, biomass can provide reliable electricity and heating. For example, Vermont's North County Hospital uses wood chips delivered from a lumber company less than a mile away to generate heat and power for its facilities.[29] Thanks to grants from the Vermont Economic Development Authority and the federal government, North County Hospital has been able to install the self-contained heat and power system, which has saved the hospital approximately $102,000 per year in electrical costs and $128,000 per year in heating fuel costs.[30] These combined systems use biomass-fueled steam boilers to both produce electricity and to heat their buildings. Another example is the Gundersen Health System hospital campus in La Crosse County, Wisconsin. Gundersen uses methane gas from a nearby landfill and wood waste from nearby lumber operations for a combined heat and power system.[31]

In the fall of 2014, Gundersen generated excess electricity, which it was able to sell to an electricity company for a profit. One of the main advantages of having an on-site boiler is that these hospitals can function independently of local power grids. If a power outage happens, these hospitals are self-reliant for at least several days, depending on the amount of fuel they have stored.[32]

Even though biomass can be a reliable source of baseload power, it still has its drawbacks. Organic materials are not very energy dense because they usually contain a large amount of water, meaning that they do not contain much potential energy when burned. Coal, for example, is a very energy dense material because it is almost completely composed of carbon. Most biomass material, including wood and grass, is bulky and water-filled. One ton of coal contains far more potential energy than one ton of wood, which means that we must burn much larger amounts of wood than of coal to generate the same amount of electricity.[33]

Biomass: The Dirty Renewable

Renewables have become romanticized as a cure-all for the pollution and climate change issues that we face today. The fact is that biomass, especially industrial-size biomass production, can have huge repercussions on the quality of our water, land, and air. As the biomass industry grows because of government support, more biomass, especially wood, is required to fuel our growing consumption. Ironically, the "green" industry of biomass has turned to logging operations specifically designed to harvest trees for electricity generation, rather than using wood waste products.[34]

A single biomass power plant can burn around 650,000 tons of lumber per year. As states try to fulfill their RPS requirements, the expanding biomass industry will need more logging to feed hungry power plants.[35] Many states have counterintuitive laws that seem to hurt the environment more than help it. For example, Ohio laws recognize native forests as "tree crops" that are eligible to be clear-cut for biomass fuel.[36] Overall, 32 other states include whole trees as a legally acceptable form of biomass fuel. As the industry continues to grow artificially fast with the government's help, biomass companies will be able to more easily engage in commercial logging in the name of "eco-friendly" energy.[37] As long as whole trees are considered an acceptable fuel under state law, whole trees will be cut and burned for the sake of the environment and climate change. If states would change their definitions under their RPS standards and other energy laws, biomass facilities would have to find more practical fuels, such as plant or animal waste.

Even though it seems nonsensical to cut and burn trees for the sake of the environment, many green energy advocates and some government agencies justify using trees as biomass fuel because it is "carbon neutral." The USDA and EPA claim that biomass is carbon neutral because the amount of carbon released by burning trees is the same amount that is absorbed when the plants regrow.[38,39] Although the logic is technically correct, this line of thinking does not take several important implications into account. In a natural setting, many plants, especially trees, grow and decay over many years, if not decades. When we burn massive amounts of trees at once, we release the plant's stored carbon into the atmosphere in a short period of time. In the long term, biomass may be carbon neutral in some circumstances, but in the short term, we increase net carbon emissions by burning trees for fuel because we release the stored carbon.[40]

Using trees for electricity is a counterproductive means of limiting carbon emissions. The Manomet Center for Conservation Sciences said, "Forest regeneration and growth will not instantaneously recapture all the carbon released as a result of using the woody material for energy generation."[41] Carbon neutrality is only true in the long run if we plant new trees at the same rate that we cut them down and burn them, but biomass facilities are being built at continually faster rates thanks to government mandates and subsidies.[42]

Much of modern energy policy seeks to encourage renewable energy in the hopes of combating climate change. Trees serve as one of the best forms of capturing and sequestering carbon, which may be one of the most effective ways of limiting human-caused climate change. Federal officials continue to pledge billions of taxpayer dollars and endless hours of lip service to combating human-caused climate change. Federal agencies continually make regulations stricter for greenhouse gas emitters for the sake of "saving" us from climate change. Despite noble intentions, the federal government's policies for biomass blatantly contradict other policies that are meant to limit climate change. The choice to subsidize and mandate the burning of whole trees is a backward way to address climate change. Although people may disagree on the causes, severity, and threats of climate change, contradictory government policies certainly do not help solve any of these potential problems.

Even though biomass is a renewable energy source, it is not nearly as green as most people assume. In fact, biomass in many cases can be as bad or worse than burning coal. Burning biomass releases more carbon into the atmosphere per unit of power produced than coal. A study by the Partnership for Policy Integrity found that burning biomass releases more carbon for every unit of electricity compared to coal. Biomass also produces more nitrogen oxides than coal per megawatt-hour.[43] The Massachusetts Environmental Energy Alliance also studied the environmental effects of biomass and found that burning biomass produces roughly one-and-a-half times more carbon monoxide and carbon dioxide than burning coal.[44]

Small biomass power plants produce more emissions per unit of energy generated than larger fossil fuel power plants. The Gainesville Renewable Energy Center (GREC) is a 100-megawatt biomass power plant in Gainesville, Florida. When the GREC was first proposed, its air permit applications showed that it was expected to produce 4.53 times more nitrogen oxides, 12 times more carbon monoxide, 5 times more particulate matter, 14.6 times more sulfur dioxide, and 3.25 times more volatile organic compounds than a much larger coal power plant.[45] In 2014, the GREC burned an average of 61,000 tons of biomass material per month, including forest residues, plant residues, wood waste from industry and residential areas, and whole trees.[46,47] Although the GREC claims that its fuel is mostly waste products, it still uses whole trees as a fuel, meaning it still has large impacts on forests. To put the amount of biomass the GREC burns into perspective, American forests yield between 1 and 4 tons of biomass material per acre.[48] If we assume that the GREC only took biomass material from forests with 2 tons of biomass per acre, it would consume about 30,500 acres, or roughly 48 square miles, of biomass monthly.[49]

In addition to trees, various kinds of grasses, especially switchgrass, have become popular in recent years as a biomass fuel. Switchgrass grows to maturity in a relatively short time, and it can thrive in a wide variety of soil types, terrains, and climates.[50] The argument that grass is carbon neutral is much more realistic than for trees because grass only needs a few weeks to a few months to fully regrow, as opposed to decades for trees. The downside to grass is that it requires huge amounts of land to grow enough grass for a biomass facility to function. Ironically, crop grasses have the potential to lead to higher carbon emissions if forestland is cleared to grow them. As mandates and subsidies artificially increase the demand for grass fuels, some forestland will likely have to be cleared to make enough space.[51] Cutting down trees releases carbon into the atmosphere, and crop grasses will not be able to store the same amount of carbon.[52]

To generate utility-scale electricity, the amount of grass needed for biomass power plants is tremendous. Researchers at Iowa State University calculated that 1 megawatt of electricity would require roughly 1,500 acres of switchgrass per year.[53] If a biomass facility wanted to generate the same amount of electricity as an average 600-megawatt coal power plant, it would need 900,000 acres of land dedicated solely to switchgrass.[54] To put that figure into perspective, we would need a field of switchgrass that was 1,400 square miles, which is slightly larger than the size of Rhode Island, to power a single average-size power plant.

Not only do grass fuels pose a potential threat to the environment, they could also threaten human welfare. The World Resources Institute concluded that increasing the worldwide use of energy crops, like grasses for fuel, directly threatens the availability and sustainability of future food supplies by increasing demand and thus increasing food prices. Higher costs for food would impact the poorest people the most. Choosing whether to grow plants for food or fuel requires a decision between significant tradeoffs. For instance, the World Resources Institute found that we would have to double the entire world's output of all plant materials if we wanted to generate just 20 percent of the world's energy demands from biomass by 2050.[55] Policy makers and members of the public must ask themselves if the environmental and monetary costs of growing biomass of fuel are worth the time and resources that could have been devoted to growing food instead.

Burning grasses, similar to burning wood, can pose several technical problems for biomass facilities in the electricity generation process. First, grass contains a relatively large amount of water, which limits the potential energy that power plants can extract by burning grass.[56] Even though energy producers can dry and pelletize grass to make it more energy dense, burning grass creates salt residues inside combustion chambers of power plants. These residues are known as "clinkers," which reduce the efficiency of the electricity generation process.[57] Most biomass power plants are designed to burn wood, so when they burn grass as fuel instead, they are ill-equipped to handle residue buildups. Energy producers can co-fire grass pellets with coal or wood in small amounts, usually less than 20 percent grass by mass, to minimize clinkers.[58] Exclusively burning grass for electricity has high economic and environmental costs, but using it as a supplemental fuel may make more sense economically and environmentally.

A Blind Eye to Biomass Pollution

Even though most biomass fuels are actually more environmentally harmful than coal in some respects, the federal government gives biomass power plants a free pass when complying with federal pollution standards. Federal agencies, especially the EPA, closely monitor all air and water pollution that might come from fossil fuel power plants. Biomass, for political reasons, is exempt from the same standards. Under the current laws and regulations, fossil fuel plants must obtain a Prevention of Significant Deterioration (PSD) permit from the EPA when they emit 100 tons or more per year of pollutants specified in the Clean Air Act, including carbon dioxide.[59] These permits are meant to ensure that air quality is protected from the side effects of electricity generation. PSD permits require companies to use emission-reducing technology and also conduct air impact evaluations.

In contrast to fossil fuel facilities, biomass power plants do not have to obtain PSD permits until they have emitted two and a half times more of the specified pollutants.[60] Biomass facilities get even more special treatment under the current regulations. They do not face restrictions for hazardous air pollutants such as hydrochloric acid, dioxins, carcinogens like benzene and formaldehyde, or heavy metals like arsenic, lead, and cadmium, but fossil fuel power plants do.[61]

The EPA gives biomass even more special treatment when defining what kind of waste biomass facilities can burn. This regulatory flexibility allows biomass facilities to burn tires, plastics, and other wastes with minimal accountability or punishment.[62] Although fossil fuel power plants are required to use the best available technology to help reduce emissions, biomass facilities are not required to use these technologies until the EPA issues them a PSD permit. With such flexibility, biomass power plants can emit harmful pollution when they burn garbage.[63]

The double standard between biomass and fossil fuels exemplifies the strange ignorance that policy makers adhere to when it comes to "renewable" energy sources. Biomass facilities are lightly regulated and highly subsidized, which means that the industry is growing much more rapidly than it otherwise would. When the government subsidizes these companies and turns a blind eye to the biomass facilities pollution, the industry has little incentive to produce clean electricity and reduce emissions. If EPA officials really want the public to believe that their purpose is to reduce pollution and protect the environment, then they should reconsider giving free passes to the biomass industry when they severely restrict the fossil fuel industry with crippling regulations.

Policy leniency or strictness comes with trade-offs, but purposefully treating two industries differently even though they produce similar environmental issues is nonsensical and results in bad outcomes for the environment and for the American people. Strict command-and-control regulations may have good intentions, but these regulations always come with costs and trade-offs that may cause additional economic or environmental harm. If our goal is to promote cleaner energy production, we should be looking for innovative solutions that produce the actual outcomes we want, rather than supporting well-intentioned but faulty policy solutions.

CHAPTER SEVEN

Geothermal

Controversial Loan Guarantees to Geothermal Companies

In the deserts of Humboldt County, Nevada, a renewable energy company called Nevada Geothermal Power (NGP) ended up in hot water after experiencing financial trouble. NGP received millions in federal loan guarantees, but a scandal has erupted after the company's first power plant became operational. In its first year of operating the Blue Mountain Geothermal Power Plant, the company racked up substantial debts because it did not produce as much electricity as the developers expected.

NGP had racked up millions of dollars of debt from private loans and government-subsidized loans. In addition to these loans, the federal government provided tens of millions of dollars in financial assistance in the form of loan guarantees and grants. Even with the government's backing, NGP still struggled to pay back its private loans because it could only produce 27 megawatts, rather than the predicted 45. As a result, the government partly guaranteed a second $98.5 million loan for NGP in 2010 to prevent the project from failing.[1]

One of the biggest supporters of NGP's loan guarantee was Nevada senator Harry Reid, who pushed for the loan guarantee program to be included in federal stimulus bills. Another supporter of the loan guarantee program was Energy Secretary Steven Chu, who supported and encouraged the federal government's financial support.

Not everyone in the federal government praised NGP's loan guarantee as highly. Representative Jim Jordan is a Republican from Ohio and then chairman of the House Oversight and Government Reform subcommittee on regulatory affairs, stimulus oversight, and government spending. Jordan said that NGP "was in danger of defaulting on its financial obligation, and the [Department of Energy's] assistance served as a de facto bailout. After receiving a taxpayer-backed $98.5 million loan guarantee, the company is still struggling." Jordan continued, saying that the loan guarantee "essentially served to prop up an already-faltering firm."[2]

Other congressmen have also criticized the dubious loan guarantees provided to renewable energy projects with the backing of taxpayer funding. Representative Darrell Issa of California, then the chair of the House Oversight and Government Reform Committee, criticized Secretary Chu, questioning why the DOE guaranteed a loan to NGP when the company had long-standing financial problems.[3]

NGP originally thought that the Blue Mountain plant would produce 50 megawatts of electricity, but the actual capacity was much lower. Because of unforeseen temperature decreases, the plant's capacity continually declined. Plant operators expected the capacity to fall to only 15 megawatts by 2020. The combination of immediate financial troubles and decreasing power output over time caused NGP to default on its private loan to EIG Global Energy Partners. In 2012, EIG took over Blue Mountain. Two years later, EIG hired AltaRock, a tech company funded by some of Silicon Valley's biggest investors, including venture capitalist Vinod Khosla. AltaRock was given the responsibility of revamping Blue Mountain's technology to bring it back up to its expected capacity.

The new owner of the project has been able to increase the plant's capacity to 40 megawatts, and plans to use revenue from this increased production to pay back outstanding loans to the DOE. As of May 2015, the project still owed $70 million to the federal agency. The CEO of AltaRock estimated that this amount would be paid back entirely by 2029. Fortunately, the AltaRock acquisition of Blue Mountain appears to be turning the financial status of the power plant around. If the technological improvements prove successful, the Department of Energy may be able to avoid the fiasco of another loan default.[4]

The Basics of Geothermal Power

Geothermal power is one of the least common sources of electricity in the United States. As of 2014, geothermal power generated 0.4 percent of U.S. electricity.[5] Even though the industry is small, geothermal power is a growing energy source, thanks to favorable government policies that are meant to boost renewable energy production. Geothermal is not as well known as other renewable energy sources, but it is becoming more well known as the renewable energy sector grows. In 2014, the United States had a total of 64 operating geothermal power plants, with 83 more projects in various stages of development.[6]

Geothermal power generates power by using the heat of the Earth. Water from the surface of the Earth trickles down through the ground and becomes superheated. Probably the most famous examples of this superheated water are the geysers of Yellowstone National Park. But we do not need a Yellowstone-size super volcano to generate electricity. If we drill deep enough, we can reach superheated water that we can pump back up to the surface. This hot water turns to steam, as it rises to the surface, which is then used to spin electricity-generating turbines.

Although the basic principles are the same, geothermal power plants come in three varieties. First, dry-steam plants pump steam directly from the ground into the turbines. This only works when the underground steam is hot enough, usually 230°C to 320°C.[7] Once the steam is used to spin the turbine, it is pumped back down into the ground. Dry-steam geothermal plants are the oldest and simplest geothermal technology. In 1904, Prince Piero Ginori Conti developed the world's first dry-steam geothermal power plant in Tuscany, Italy.[8] Today, the world's largest complex of dry-steam geothermal power plants is located in northern California at The Geysers.[9]

The second type of geothermal electricity generation are flash-steam geothermal power plants, which pump superheated water into a tank at the surface. Even though the water is usually 200°C to 320°C, it is still a liquid because of the immense pressure underground. In the tank, the geothermal water is exposed to the pressure at the surface and the superheated water vaporizes into steam, which is used to spin a turbine.[10] After the steam is used, it is pumped back into the ground.

Third, binary-cycle geothermal plants produce electricity in a less direct method than dry-steam or flash-steam. Geothermal developers pump the superheated water into a heat exchanger where the heat is transferred to a secondary fluid. The secondary fluid has a lower boiling point than water, so it is easier to vaporize. The secondary fluid then spins the turbines, instead of the geothermal water. Binary-cycle systems can use geothermal water with temperatures ranging from 120°C to 190°C.[11] The geothermal water is then pumped back into the ground, while the secondary fluid is recycled in a closed loop.

Federal Incentives for Geothermal Power

The federal government first enacted policies for the geothermal industry beginning in the 1970s. Congress passed the Geothermal Steam Act in 1970, which granted the Bureau of Land Management (BLM) the power to lease public lands to geothermal developers.[12] A few years later in 1974, Congress passed the Geothermal Research, Development, and Demonstration Act. Under this act, the Geothermal Loan Guarantee Program provided over $136 million in loan guarantees to geothermal projects between 1974 and 1982. The program expired in 1982.[13,14]

In 1978, Congress passed the Public Utility Regulatory Policy Act (PURPA), which ended the monopoly that utilities had on owning and running electricity-generating facilities. PURPA required utilities to purchase power from independent power providers (IPPs), as long as it was cheaper than electricity from utility-owned power plants.[15] The act also allowed the Federal Energy Regulatory Commission (FERC) to require utilities to transmit electricity generated by IPPs.[16] When utilities are required to transmit electricity from other companies, IPPs are able to free-ride on the utility's transmission infrastructure without having to build their own expensive transmission lines.[17] The 1970s proved beneficial for renewable energy sources because PURPA and the global oil crisis gave renewables an advantage in the energy market.[18] The geothermal industry continued to grow dramatically through the 1980s.[19]

The geothermal industry's momentum under PURPA started to fizzle in the 1990s for at least two reasons. First, natural gas became significantly cheaper, which made it an attractive alternative to more expensive options like geothermal. Second, electricity markets became more competitive when regulatory changes shifted energy markets. Renewable energy sources like geothermal lost their edge in the market.[20] Geothermal power would have to wait about a decade before new federal policies altered markets to make it more viable.[21]

In 2001, the Department of the Interior and other federal agencies held a conference with the goal to encourage renewable energy development on public land. At the conference, the BLM developed a National Energy Policy Implementation Plan, which included many action items to help make it easier to develop geothermal power on BLM lands. The BLM lifted environmental regulations that had raised the costs of geothermal development previously.[22]

In 2004, the American Jobs Creation Act made geothermal power and other renewable energy sources eligible for the federal production tax credit (PTC) for the first five years of production, which was later extended to 10 years by further legislation.[23] The PTC provides a subsidy of 2.3 cents for every kilowatt-hour of electricity from geothermal sources for the first 10 years of a power plant's operations. This amount is the same as for wind power. Although the PTC was originally enacted in 1992, geothermal power was not included until Congress passed the 2004 American Jobs Creation Act. Congress has allowed the PTC to expire and then renewed it several times. Just a few of these renewals include the American Recovery and Reinvestment Act of 2009, the American Taxpayer Relief Act of 2012, the Tax Increase Prevention Act of 2014, and the Consolidated Appropriations Act of 2016. The Consolidated Appropriations Act of 2016 allowed geothermal developers to receive PTC money if they began construction on their projects before December 31, 2016.[24]

Whenever a PTC expiration has loomed in the past, investments in renewable energy sources, including geothermal, have plummeted. For example, when the PTC was set to expire again in 2014, the American Council on Renewable Energy predicted that the number of new geothermal projects would decrease, largely because of the PTC expiration.[25] The geothermal industry experienced little growth in 2014, and the Geothermal Energy Association attributed the uncertainty regarding the expiration and possible renewal of the PTC as a main cause.[26]

In addition to PTC money, several other federal acts have helped subsidize the geothermal industry. The Energy Policy Act of 2005 changed the rules for leases to federal land and how much money the federal government extracts from geothermal operations, making investments in the geothermal industry more economical.[27] The American Recovery and Reinvestment Act of 2009 allocated $368.2 million dollars to the Geothermal Technologies Office. The office divided the funds to 148 geothermal projects throughout the country.[28] The decade of subsidies and greater access to public land has helped the geothermal industry to grow rapidly. From 2005 to 2014, the amount of electricity generated from geothermal increased by 13 percent.[29] In 2013 alone, the federal government spent $312 million in direct expenditures to the geothermal industry, constituting about 90 percent of all government support to the industry that year.[30]

The Business Energy Investment Tax Credit (ITC) is another federal program that helps boost the geothermal industry. If a geothermal project is completed by the end of 2016, the program gives the project a 10 percent tax credit on all equipment that the developers purchased for that project. The tax credit does not include money spent on transmission equipment, however.[31]

Other than subsidies, the federal government also provides a system of loan guarantees to help the geothermal industry grow faster than it normally would under a truly competitive market. Loan guarantees are less common than tax credits and subsidies. In fiscal year 2010, for example, the geothermal industry received $13 million in loan guarantees from the DOE.[32] In 2011, Nevada Geothermal Power Company, Inc. was given a loan guarantee of $99 million. During that same year, Ormat Nevada, Inc. received a $350 million loan guarantee, and US Geothermal, Inc. received $97 million.

Some states also enacted their own policies to help promote the geothermal industry at nearly the same time as the federal government. California initiated its Geothermal Grant and Loan Program in 1980 to give geothermal developers monetary aid.[33] The program makes these grants and loans from revenues that the federal government pays to California for existing geothermal facilities on federal lands in the state.[34]

In addition to monetary support, many states mandate renewable energy, and geothermal energy counts toward fulfilling many of those mandates. Renewable portfolio standards (RPS) require states to meet specified amounts of renewable-generated electricity consumption, production, or sales.[35] No state has a specific RPS requirement for geothermal power, but most states allow geothermal to qualify for general RPS requirements.[36] In some places, geothermal power companies have benefited from RPS. To fulfill Nevada's RPS requirements, Nevada Power, a utility company, specifically buys power from three geothermal power plants in the state to help meet that requirement.[37]

Difficulties in Producing Geothermal Power

Geothermal power faces geographic and financial barriers that can make it one of the most complex ways that we generate electricity. Geothermal power is one of the most geographically dependent energy sources, which means that geothermal power plants are viable in only a few places on the globe. Second, developers often face uncertainty about whether a given spot will be viable or profitable until after they have already spent significant amounts of money to drill. Third, geothermal power plants need large upfront capital investments to make them work, a problem that is compounded when mixed with the geographical dependence and uncertainty.

An adage from the real estate business that it's all about location, location, location, is descriptive of the problem geothermal energy faces. In theory, geothermal resources can be found at every point on Earth, but in practical terms, it is much more difficult to locate viable locations for geothermal energy production. If someone dug deep enough, they would eventually reach a hot enough place that could be used to boil water to power a steam turbine. Some places, however, are much easier to access geothermal resources. In the United States, most accessible geothermal resources are in the western part of the country, where volcanism is more common. For geothermal power to be viable for commercial production, a location needs to have high underground temperatures that are relatively close to the surface, as well as access to underground or surface water.[38] At the end of 2013, the 124 geothermal power projects under development or planning were mainly located in the western states of Nevada, California, and New Mexico.[39]

Although geography is a key part of geothermal development, choosing the right location to drill for geothermal resources may not be enough. The exploration and drilling phases of a geothermal project are expensive, and without a high amount of certainty, many investors do not want to put money into a project that may not be profitable in the end. The uncertainty of finding a viable location for a geothermal project makes it difficult for many geothermal projects to find funding on the market.[40]

Geothermal developers can spend millions of dollars when trying to explore and drill for geothermal resources before they realize they have to abandon the project.[41] The International Finance Corporation looked at over 2,500 geothermal wells and found that 78 percent of well drillings were successful on average. These success rates, however, varied depending on the location. About one-third of the geothermal fields in the study had drilling success rates of 60 percent or lower. Thirteen percent of fields had success rates of 90 percent or higher.[42]

Drilling for geothermal resources is more difficult and more intensive than drilling for oil. Although the basic principles for drilling are the same, drilling for oil is cheaper and less risky. The average success rate for geothermal drilling is about 78 percent, but the average success rate for oil wells is about 90 percent.[43] Geothermal drilling is more difficult and costly than drilling for oil for a few reasons. First, oil reserves are most often located in sedimentary layers, but geothermal reserves are usually located in metamorphic and igneous rocks. These types of rocks are usually harder, more abrasive, and more corrosive, which requires a larger investment in tougher equipment.[44] Researchers at the Massachusetts Institute of Technology found that drilling a geothermal well costs as least twice as much as an oil or gas well of the same depth.[45] Second, geothermal production requires more exploratory drilling than oil or natural gas because more data exist for oil and gas. Third, the software used to evaluate and model subsurface resources for oil and gas resources is more robust and has existed longer. The same software does not work as well for analyzing geothermal resources.[46] With all these factors combined, the profit-to-investment ratio for geothermal power is lower than for oil and gas, meaning that it is often more profitable to drill for oil or natural gas rather than geothermal resources.[47]

The initial capital costs can be a large barrier to many would-be geothermal developers, and without government subsidies and loan guarantees, the upfront costs would be too much for most developers. Compared to other mainstream energy sources, geothermal power plants have some of the highest capital costs. The Energy Information Administration found that the capital costs of a 50-megawatt flash geothermal power plant were $6,243 per kilowatt. For reference, the capital costs for a larger 85-megawatt combustion natural gas plant were just $973 per kilowatt.[48] The International Energy Agency has stated that geothermal's high capital costs are one of the main deterrents to additional investment in the industry.[49]

Government subsidies and loan guarantees make it easier for developers to overcome the high risks and startup costs for geothermal power development. These subsidies and loan guarantees, however, may end up being paid for by taxpayers. With such support, geothermal power companies face moral hazard. When developers do not have to worry about failure, they worry less about making poor investments and may make riskier decisions than they would if their own money was on the line. Without subsidies or loan guarantees, geothermal developers would have a stronger incentive to minimize potential losses by avoiding unnecessary risk.

As technology improves, the risks and the costs with geothermal development will most likely decrease. Over the past half decade, success rates for exploratory drilling have been improving due to advancements in geothermal surveys and other technologies.[50] The federal government has also been investing in research and development for advancements in the industry. In 2011, the Department of Energy provided $70 million to improve geothermal resource evaluation, exploration, drilling, and engineering.[51] By improving technology, reducing risk, and increasing cost-effectiveness, research and development could mean that geothermal developers would need fewer subsidies and loan guarantees in the future.

One of the primary benefits of geothermal power is that it can provide baseload power that is reliable and consistent. Another benefit is that geothermal has low operations and maintenance costs once developers build a working power plant. After passing the initial hurdle of exploration, drilling, and construction, geothermal power plants are long-lasting and environmentally friendly electricity sources. The high upfront costs and uncertainty with the initial phases of geothermal development are an undeniable challenge, but that is only one part of the equation. The main advantages of viable geothermal plants is that they have a long lifetime and produce consistent, reliable electricity with zero fuel costs.

Compared to fossil fuels or even other renewables, geothermal power is one of the most environmentally friendly electricity sources. It does, however, have some environmental costs, even if those costs are relatively small. First, geothermal fluids can contain toxic chemicals, like hydrogen sulfide, arsenic, boron, mercury, lead, and aluminum.[52] If not managed properly, these chemicals can contaminate local environments and water sources.[53] To avoid contamination, many geothermal developers inject fluids back into the ground. The geothermal wells are cased to make sure that geothermal fluid does not mix with groundwater.[54] In cases where surface runoff occurs, geothermal developers place geothermal fluids in impervious holding ponds.[55] These measures seem to have been successful as there has not been a single reported instance of geothermal-caused water contamination in the United States.[56]

Other than water contamination, some geothermal power plants use large amounts of water, which may be a concern for water supplies in arid areas. The amount of water used in geothermal electricity production varies widely depending on which type of technology is used. For example, geothermal plants with cooling towers can be one of the most water intensive renewable energy sources. In contrast, geothermal plants with dry cooling systems can be one of the most water efficient renewable energy sources.[57] The sources of water for geothermal power plants can come from local freshwater sources, recycled geothermal fluid, or even sewage water.[58,59]

Second, geothermal power plants can emit a minor amount of pollution. Flash and dry-steam plants directly use geothermal fluid, which contains dissolved gases that are released into the atmosphere when brought to the surface.[60] These gases include nitrogen oxide, sulfur dioxide, particulate matter, and small amounts of carbon dioxide.[61] Compared to coal power plants, flash and dry-steam geothermal power plants emit less than 1 percent of the nitrous oxide, 1 percent of the sulfur dioxide, and about 5 percent of the carbon dioxide emitted by a coal-fired plant of the same capacity.[62] Binary geothermal plants vaporize a secondary fluid to produce power instead of using steam from geothermal fluids, and produce virtually no emissions.[63]

Third, geothermal power plants can cause land to sink, also called subsidence, because water is being taken out of the ground and thus reducing underground pressure. The shifting of the ground from water removal may induce seismic activity. Scientists in southern California found a strong correlation between seismic activity and how much fluid geothermal power plants were extracting and injecting.[64] In northern California, other scientists have found that all the earthquakes that have occurred at the geothermal power complex at The Geysers since 1975 have been caused by geothermal energy production.[65] Geothermal developers can reduce seismic activity by keeping their drilling and fracturing relatively small in any one location.[66] Although seismic activity cannot be ignored, no earthquake that has been attributed to geothermal power has caused damage.

CHAPTER EIGHT

Hydropower

Misadventures with Micro-Hydropower

Logan, Utah, is a university town nestled in the mountains of northern Utah. In 2008, city officials realized that they needed to install larger municipal water pipes for the rapidly growing population. The city's drinking water comes from Dewitt Natural Spring in the mountains nearby. City officials and engineers chose to replace the original municipal water pipes from the spring with larger ones to allow more water to reach residents. The problem with the larger pipes, however, was that the water pressure in the new pipes was too high.

The city's engineers saw that pressure-regulating valves were not a permanent solution to the problem. Logan's assistant engineer, Lance Houser, came up with the idea to reduce the pressure with a hydroelectric turbine that would also produce clean electricity with very little up-front cost. Houser and the other engineers moved ahead with the plan to install a 200-kilowatt electrical turbine in the city's water pipes, which would not only reduce water pressure to acceptable levels, but also would power roughly 185 homes.[1]

Despite the ingenious solution to the city's water problem, the project became a complicated and expensive mess because of a maze of federal regulations that the city engineers did not foresee. For nearly a year, Logan city officials muddled through the federal permitting process. Even though the pipes had existed in the same place for decades, the Federal Energy Regulatory Commission (FERC) required the Logan project to abide by a litany of federal laws that were designed to apply to large dams with significant environmental impacts. Among the regulations, the micro-hydropower project had to conduct an environmental assessment, assess water quality, document impacts to wildlife, and catalog any potential effects the turbine installation would have on historical structures. All of this was required despite the fact that the project would involve nothing more than placing a turbine within an already existing pipeline.

Houser and his team found that one of the biggest regulatory hurdles to their project was complying with the National Environmental Policy Act (NEPA). Congress passed NEPA to make sure that any projects overseen by federal agencies consider their environmental impacts before they move forward. Project leaders, whether a government agency or private company overseen by an agency, must fill out extensive paperwork on the potential environmental impacts from their project, as well as alternative plans that pose fewer environmental impacts.

FERC can "categorically exclude" a project if officials do not expect it to have any significant environmental impacts. If a project does not get the exclusion, FERC can require project leaders to conduct an environmental assessment (EA) to determine the impacts of their project. If FERC decides that the project will not have a significant impact, then it can move forward, but if there are potential significant impacts, FERC may require an environmental impact statement (EIS). An EIS is more extensive in terms of scope and time than an EA. An EIS also allows outside parties the opportunity to provide input, which often makes the process lengthier.

FERC may grant exemptions to small-scale municipal hydropower projects from certain licensing requirements. These exemptions are only granted if the project is under 40 megawatts, if the existing pipeline or canal was not originally built for power generation, and if the project is not located on federal lands. Logan officials applied for and were granted a conduit exemption in January 2011, but Houser said, "the irony of the whole thing is that you get the exemption but all it saves you is about three to four months."[2]

The exemption, however, only applied to the formal EA, and FERC required Logan's officials to provide a draft preliminary EA. This requirement made little sense because the water used in the turbine comes from Dewitt Springs, not a river, and the project would only affect an existing pipeline. Complying with all of FERC's regulations alone added roughly $400,000 to the project. The project's total cost amounted to nearly three million dollars and took four years. For comparison, a similar project in Canada would only have total costs between $225,000 and $375,000.

Logan's experience with micro-hydropower is not unique. Over the past 40 years, the costs of complying with regulations have risen from 5 percent to 25 percent of total project costs for conduit-based hydropower like Logan's. Policies like the Federal Water Power Act, the National Historic

Preservation Act, and the Energy Policy Act can unintentionally discourage hydropower projects by raising the costs of obtaining a hydropower license. Houser explained that because of "the cost of the permitting headache and the nightmare and the frustration of the process, there is no economic benefit to doing a project that size again."[3]

Hydropower in the United States

Hydropower is one of the oldest and most important electricity sources in American history. For over one hundred years, Americans have been enjoying reliable electricity thanks to publicly and privately owned dams fitted with hydroelectric turbines. Some of America's most famous landmarks are hydroelectric dams, such as Hoover Dam near Las Vegas, Nevada.

Dams, however, have been controversial for nearly as long as they have been generating electricity. The Glen Canyon Dam in Arizona and the Hetch Hetchy Project in California's Yosemite National Park sparked such bitter opposition when they were built that people are still fighting to remove them many decades after they were completed. The Glen Canyon Dam was so controversial that author Edward Abbey wrote the novel *The Monkey Wrench Gang* in 1975 to protest the dam. The book advocated for sabotage against the dam that would spoil the beautiful Colorado River and its unique canyons.[4] In 2012, San Francisco on Proposition F, which would have required a study on draining Hetch Hetchy Reservoir and finding a new water delivery system for the San Francisco Bay area was on the ballot, and the proponents of the measure focused heavily on the impacts the dam had caused on the ecosystem. The proposition failed with 77 percent of voters saying no, but people continue to file lawsuits to remove the Hetch Hetchy Project.[5]

Large-scale hydropower has been controversial in some cases throughout American history, but large dams are not the only sources of hydropower. Small- and micro-scale hydropower offers potential solutions to our energy needs, especially as people become more concerned with air emissions.

All hydropower, whether large or small, works in essentially the same way. Flowing water is used to spin the blades of a turbine, which in turn powers an electric generator. Using this basic principle, hydropower can come in three distinct forms. First, impoundment hydropower plants use a dam to hold water. When the water is released through the dam, the flowing water spins the hydroelectric turbines. Electricity from the generator is then transmitted to the grid where people can use it. Impoundment plants are the most common way to generate hydropower.

Dams, however, are not necessary to generate hydroelectric power. Diversion power plants use already flowing water to power an electric generator. Diversion power is also referred to as run-of-river or conduit hydropower because water is usually siphoned off the main part of a river into a canals or pipes where it spins the turbines. The third kind of hydropower plants are known as pumped storage. These power plants work like a battery because they use excess electricity to pump water uphill to a reservoir. When electricity demand goes up, the pumped storage reservoir can release water back downhill, where it spins a hydroelectric turbine. Pumped storage power plants function almost exactly the same as impoundment dams, except that plant operators pump the water to a reservoir rather than building a dam to limit the flow of an existing river.[6]

Many environmental advocates do not want to allow the construction of any new large dams because of the negative environmental impacts and damage they can do to scenic areas. Even if dams were not controversial, most of the best sites already have dams built on them, or preservation laws protect them from construction. The combination of public outcry and a limited number of viable locations means that new large-scale hydroelectric dams are not likely to be constructed in the future.[7]

Even without new dams, we can still use existing hydroelectric dams to generate more power than we have in the past. Energy producers can improve their hydroelectric turbine's efficiency through a process known as uprating. If all the large-scale hydropower facilities in the United States underwent uprating, our hydroelectric production could increase between 8 to 50 percent.[8] This would occur with minimal or no additional environmental impacts because uprating involves completely replacing outdated and inefficient turbines or simply installing more accurate computer controls for gates, switches, and monitoring equipment.[9]

America's Widespread Potential for Small-Scale Hydropower

Although new dam construction in the United States is unlikely, we do not need new dams to increase the number of hydroelectric facilities. If the United States' 54,000 non-powered dams were retrofitted with hydroelectric turbines, we could potentially add up to 12.1 gigawatts of new electricity capacity across the country.[10] These non-powered dams are generally smaller than typical hydropower facilities and are found in all 50 states.

Many people, especially in the Pacific Northwest, are currently advocating for the removal of many non-powered dams that are harming river ecosystems and blocking fish migration patterns. Removing some dams may be environmentally beneficial, but in more densely populated places, retrofitting non-powered dams may be a way to produce electricity with little to no additional environmental impacts. Many of these non-powered dams have existed for decades, so these dams have already disturbed the area's ecology. In many cases, the environmental impacts of these dams are irreversible without completely removing the dam. Dam removal is unlikely in most circumstances because dams serve multiple purposes, including water storage and flood prevention. Because most non-powered dams are not likely to be removed and have already impacted the environment, it may be prudent to retrofit them with turbines when it is economical.

To expand our use of hydropower, however, we do not even need dams at all. Conduit-based hydropower can generate electricity from turbines installed in man-made waterways, like aqueducts, tunnels, and pipes.[11] These pipes are not in short supply. Nearly every city in the United States has a municipal water system, and most places in the western United States have canals and irrigation systems. Flowing water is almost everywhere, and we could tap that movement to generate electricity. By repurposing existing water infrastructure, water conduits could provide a viable green energy solution that many Americans want.

Conduit-based hydropower could provide more electricity, but like all energy sources, conduit hydro comes with trade-offs. The overall output from conduit systems depends on the efficiency of the turbine, as well as the distance, speed, and volume of flowing water at a specific location. Conduit-based projects do not produce as much electricity as other forms of hydropower because the water flow in many conduits is not as consistent or fast as dams.[12] Even with its limitations, conduit-based hydropower could still provide a significant amount of new hydroelectric resources. For example, California could potentially produce up to 1,122 more gigawatt-hours of conduit hydropower per year, while Oregon could potentially produce an additional 9,040 gigawatt-hours per year, and Massachusetts an additional 39.5 gigawatt-hours per year.[13]

One of the main reasons that conduit-based hydropower is not very common is because there are few existing projects that potential developers can imitate. Only a small number of conduit hydropower plants exist, so most developers do not have much or any experience with the technology. Investors are usually wary of new or unproven technologies because they want to base their investments on operational track records that have shown solid results.[14] Conduit-based projects are even more uncertain to most investors because financial data are so limited. The Federal Energy Regulatory Commission does not require these projects to state their cost information on their applications or application exemptions because they are so small.[15] Not only is financial data limited, but any type of data for conduit-based hydropower systems is difficult to obtain because there are such few systems in the United States. No government agency or academic organization has conducted comprehensive national assessments of the energy potential from conduit-based hydropower.[16]

Publicly owned utilities own many of the water pipelines in the United States, which makes taking advantage of conduit-based hydropower even more complicated. The members of elected boards for publicly owned water utilities often fear two things. One, they fear the public's response if they undertake an unpopular project, which may end up costing the board members their jobs. Second, they fear that these hydropower projects will not be economical, subsequently resulting in higher electricity prices. Making anything more expensive is often a death sentence for an elected official's job. Even if water utilities had the desire to move forward with a conduit-based hydropower project, they are constrained by the fact that they have limited budgets. Decision makers at utility providers are usually averse to using their limited budgets on projects that do not help them achieve their core functions.[17] Officials at water utilities are required to provide access to clean water, not to generate electricity, so conduit-based hydropower projects offer potential benefits that are outside of the scope of the utility's primary interests.

Regardless of its size, hydropower is one of the most reliable sources of renewable energy, and one of the most flexible sources of power to meet electricity demand. Unlike most other electricity sources, hydropower can provide both baseload power and peak power. In general, hydropower runs consistently and does not require much maintenance, which makes it an ideal source of baseload power. In many parts of the country, especially the Pacific Northwest and parts of the South, millions of people rely on hydropower to generate the majority of their electricity. It can provide baseload power because most sources of water are always flowing. For large and small dams, reservoirs hold a fairly constant supply of water. Dam operators can let out a consistent amount of water through the turbines to provide a stable, predictable amount of electricity. During the day, the electricity demand fluctuates, so grid managers and dam operators work together to let additional water through the turbines to provide the peak amounts of electricity when demand is highest.[18]

Hydropower, both large and small scale, is one of the most efficient forms of renewable energy. According to the Energy Information Administration (EIA), hydroelectric power plants in the United States have an average capacity factor of 39.8 percent.[19] This number may seem low compared to other baseload energy sources, which are usually around 75 percent or higher, but hydroelectric dams serve the dual purposes of both generating electricity and storing water for future use. If dams were to output as much electricity as possible, they may let out too much water that could be used for other important functions, like drinking water and agriculture. For example, Hoover Dam could achieve a higher capacity factor if it let out more water, but the water in Lake Mead behind the dam is highly valued elsewhere. That water is used for municipal water supplies and agriculture in the deserts of the Southwest. Millions of people rely on the electricity and the water that comes from Hoover Dam, so dam managers must balance the needs for electricity and water in one of the fastest growing regions of the country. Other baseload power sources, like coal and nuclear plants, generally have higher capacity factors than large hydroelectric facilities because their sole purpose is to generate electricity.

Small and micro hydroelectric plants have capacity factors similar to large-scale hydropower. On average, the capacity factor estimates for small- and micro-hydropower ranges from roughly 20 to 60 percent, depending the unique circumstances surrounding each hydro facility. For many small-scale hydropower facilities, their capacity factor is significantly higher than the estimates for both wind and solar power, even though these two energy sources are much more politically popular. Most estimates for wind power's capacity factor range from 30 to 35 percent, and estimates for photovoltaic solar power's capacity factor are rarely higher than 20 percent. Large-scale hydropower has a capacity factor ranging between 30 and 50 percent.[20] In some cases, small-scale hydropower can have a higher capacity factor than large hydro when water flow is more consistent. Capacity factors for any type of hydropower plant can vary widely because dams have to balance the responsibilities of electricity generation and water storage. In particularly dry areas, dam managers may need to limit the flow of water out of dams to save water for later, which would decrease the amount and consistency of water available for electricity generation. Hydropower also serves as a way to meet peak power demand, so grid managers may keep electrical output artificially low so that they can change the output to meet electricity demand exactly.

Although reliability is crucial to an electricity source, the costs of hydropower can also determine whether a project will be successful. The monetary costs of hydropower projects are largely based on their location. Projects that happen to be far from existing transmission networks can be significantly more expensive than projects that can be easily linked into a local power grid. Although installing a large-scale hydropower project can range from $1,000 to $3,500 per kilowatt, the largest scale projects can cost much more.[21] Retrofitting non-powered dams can be much more affordable than building new ones, costing as low as $500 per kilowatt under ideal circumstances. The Energy Information Administration estimated capital costs of approximately $3,000 to $6,500 for coal plants, $5,500 for nuclear power, and $4,000 to $5,000 for solar.[22]

As Americans and their elected officials continue to push for more renewable energy, the most viable way to boost our renewable electricity production may be through hydropower. Retrofitting non-powered dams with turbines is a viable option that policy makers and the general public often overlook. Even though the United States has tens of thousands of non-powered dams, making a blanket statement that we should retrofit every non-powered dam is unwise. The relatively large number of non-powered dams does not tell us anything about the unique circumstances or economic viability of any individual dam. Regardless of the exact number, thousands of non-powered dams could provide us with reliable electricity.

Non-powered dams are located throughout the United States, and hundreds of them have a potential capacity of greater than 1 megawatt. All of the lower 48 states have the potential to convert existing non-powered dams into hydroelectric plants. Most states have at least one non-powered dam with a potential capacity of at least 1 megawatt, and some states have dozens of opportunities to take advantage of this electricity source. Arkansas, Illinois, and Kentucky have some of the most potential to take advantage of their non-powered dams. Delaware, Nebraska, and South Dakota have some of the lowest potential.[23]

The DOE's Oak Ridge National Laboratory found that retrofitting non-powered dams could provide an additional 45 terawatt-hours of electricity per year. This much new electricity would constitute a 15 percent increase in existing hydropower output. This huge boost to potential hydropower production comes from the top 100 non-powered dams. If the top 10 non-powered dams alone were retrofitted, they could add approximately 3 gigawatts of new capacity.[24] For comparison, Hoover Dam has a capacity of 2 gigawatts and produces 4.5 million megawatt-hours every year on average, which produces electricity for 8 million people in the Southwest.[25]

Some argue that installing turbines on non-powered dams may harm fish populations. Turbines in hydropower plants can pose a danger when fish swim into the spinning blades.[26] In addition to direct contact with the turbine blades, turbine-caused changes in water pressure or an oversaturation of nitrogen can also kill fish.[27] The good news is that hydroelectric power plant operators can mitigate fish mortality by choosing which type of turbine to use. Francis and Kaplan turbines, for example, are two common types of turbine designs. Each of these turbines allows different species of fish to survive at higher rates. Developers can choose the type of turbine and other mitigation techniques to make sure that fish and other aquatic wildlife are protected as much as possible. As with all electricity sources, there will be trade-offs between costs and benefits. Retrofitting non-powered dams would produce clean, reliable electricity, but it also may raise fish mortality to some degree. These trade-offs will continue to be debated in the market and in the political sphere.

Regulatory Barriers to Small Hydropower

One of the biggest barriers to taking advantage of non-powered dams is the heavy regulatory burden that makes the licensing process for retrofitting non-powered dams costly and unappealing. The complex and time-intensive process burdens developers. Even though many politicians and environmentalists want to develop renewable energy sources, more conspicuous forms of renewable energy, especially wind and solar, receive billions in subsidies from the federal government. Although small-scale hydropower is one of the most reliable forms of renewable energy, a tangled maze of federal laws and regulations deters many would-be hydropower developers. The combination of burdensome regulations and large subsidies for other renewable energy sources has made it hard for hydropower to compete in the energy market. Until 2005, hydropower did not receive the production tax credit (PTC).[28] Qualified hydropower projects only receive half the amount of PTC subsidies that wind power, biomass, and geothermal do.[29]

If a landowner wanted to install a small-scale hydroelectric facility on her property, she would have to gain permission from FERC. FERC can give exemptions for residential-size hydropower projects, but the process is costly and time consuming. To apply for FERC's typical exemption for a small hydropower system, landowners or potential developers usually need to hire a set of consultants, including lawyers, surveyors, and engineers, who know how to navigate the regulatory maze. Once the consultants complete the application for an exemption, the entire document may be roughly 100 pages long. These applications may include explanatory text, as well as any other information that FERC officials may find necessary, such as diagrams, maps, and appendices. Completing the entire exemption application can take several months and cost between $10,000 and $30,000. These costs often amount to more than the cost of the actual hydroelectric equipment.[30]

Uncertainty in the permitting process has become one of the largest barriers to hydropower development. When the regulatory regime introduces additional uncertainty into the development process, the whole hydropower industry suffers. Hydropower is different than most other electricity sources because most other sources have more certainty regarding the development schedule, permitting approval, and the availability of a power purchase agreement (PPA).

Because most hydropower projects are dependent on government approval, hydropower developers cannot control how quickly the development schedule and permit approval will be completed. When firm development schedules and permit approvals are not set, hydropower developers cannot set a guaranteed project time line. Without a firm project time line, hydropower developers have a difficult time establishing a PPA contract with a grid operator or utility provider. Without the assurance of a PPA, lenders are much less likely to make long-term investments with hydropower developers, who, in turn, are unlikely to make long-term capital investments. The Governor's Energy Office in Maine found that the state's hydropower developers said financing hydropower projects is virtually impossible without a PPA.[31]

According to Oak Ridge National Laboratory, over the past 30 years, regulatory costs have increased from 5 percent of a hydropower project's total cost to as much as 25 to 30 percent.[32] Conduit-based hydropower faces similar regulatory costs as conventional hydropower. Regulations for conduit-based hydropower include permitting, licensing, postlicensing, relicensing, environmental mandates, and engineering assessments to make sure that existing water delivery functions remain unharmed.[33]

Despite the regulations that stifle hydropower development, state legislatures have established state-level mandates in their renewable portfolio standards (RPS) to encourage hydropower. Each state has a different definition of which types of renewables are acceptable and how much electricity each renewable energy source should generate. Hydropower is a renewable energy source by any reasonable definition, but some states passively discourage hydropower because legislators excluded many types of hydropower in their RPS laws.[34] Each state discriminates on which types of hydropower projects qualify for RPS based on its capacity, how old it is, or what technology it uses. In many cases, existing hydropower facilities are too old or too large to qualify under a state's RPS mandates as they are currently written.[35]

For the states that do include hydropower in their RPS mandates, some include any kind of hydropower development, while others only include new construction. For example, North Dakota's RPS allows for any kind of hydropower development to count toward its renewable energy mandates. New Mexico, on the other hand, only accepts hydroelectric facilities built after July 2007. By only accepting new construction, New Mexico's policy fails to consider the environmental costs of encouraging new dams. Illinois, Michigan, and Missouri take a more middle-of-the-road approach to their hydropower RPS policies. These states' RPS requirements only include hydropower that does not require new dam construction or significantly expanding existing dams.[36]

Despite seemingly noble intentions, laws that were meant to preserve environmental quality, human health, and historic structures can actually stifle clean energy development with minimal, if any, benefits. The one-size-fits-all approach to many federal laws makes them unhelpful in unique circumstances, especially when it comes to retrofitting non-powered dams or installing conduit-based systems. These well-intentioned policies include the Federal Water Power Act, the National Historic Preservation Act, the National Environmental Policy Act, the Clean Water Act, and the Energy Policy Act. Each of these policies raises the cost of all hydropower projects, which discourages the growth of one of the most viable renewable energy sources.

The Federal Power Act of 1920, also known as the Federal Water Power Act, established the Federal Power Commission, the predecessor to today's Federal Energy Regulatory Commission (FERC). FERC has the power to issue licenses to nonfederal hydropower projects that affect navigable waters, occupy federal lands, affect interstate commerce, or use water from government-operated dams.[37] Through the Federal Power Act, hydropower developers must obtain a hydroelectric license from FERC, even for retrofitting existing dams and conduits.

FERC must obey Section 106 of the National Historic Preservation Act, which requires FERC officials to examine how proposed projects will affect any historic property.[38] Although intended to protect the nation's heritage, the National Historic Preservation Act can impede hydroelectric projects. Small-scale hydropower projects must file the appropriate paperwork regarding the impacts of a hydropower project on nearby historical structures, even if the project will clearly have no impact on historical structures.

The National Environmental Policy Act (NEPA), which requires all federal agencies to analyze the environmental impacts of a government project or a project on government property, is one of the environmental laws with the widest reach. Under NEPA, FERC must draft an environmental assessment (EA) or environmental impact statement (EIS) for hydroelectric projects. These documents can be hundreds of pages long and take months to complete. These documents also require FERC, as well as all agencies, to suggest alternative plans that will be less environmentally harmful than the initially proposed plan. The legislators who helped to pass NEPA wanted to make sure that all environmental implications of a government project were considered before agencies went ahead with a project. For hydroelectric projects, the lengthy process of NEPA compliance can discourage developers or increase the costs.[39] Although retrofitting existing dams or installing conduit systems has virtually no environmental impact in most cases, FERC still requires potential developers to prepare an EA or EIS.

FERC generally uses the water quality standards from the Clean Water Act as a condition for project certification. Developers must find ways to meet minimum water quality standards, or their projects will not be certified. FERC also consults with both the U.S. Fish and Wildlife Service and National Marine Fisheries Service before issuing a hydroelectric power license to make sure that hydropower projects will not harm any listed species or their habitat under the Endangered Species Act.[40] New dams are not the only hydropower projects that are subject to the Clean Water Act and the Endangered Species Act. Developers who want to retrofit nonpowered dams or install conduit systems must comply with designated water quality standards under the Clean Water Act and investigate potential impacts on endangered species under the Endangered Species Act, despite those impacts having occurred in most cases.

Despite all the laws and regulations that slow hydropower development and add to its costs, the federal government also subsidizes the expansion of hydropower at existing dams. Under Section 242 of the Energy Policy Act of 2005, the Department of Energy (DOE) was given the authority to grant subsidies to expand hydropower energy development at existing dams and impoundments. Although the law created this subsidy program, the DOE did not receive any funding for the program until nine years after the law was passed. The Fiscal Year 2014 Omnibus Appropriations bill allocated $3.6 million to the Section 242 program.[41] Congress appropriated $3.96 million for the Section 242 program in FY 2015 and another $3.5 million in FY 2016.[42] These appropriations allow qualified hydroelectric projects to receive a guaranteed payment of 2.3 cents per kilowatt-hour for up to 10,000 kilowatt-hours a year, depending on whether Congress continues to appropriate funds to the program.[43,44]

Complying with the various laws is just the first step in allowing hydroelectric projects to move forward. The convoluted licensing process provides another barrier for hydropower developers to overcome. Not only is the licensing process for hydropower, especially small-scale hydropower, unnecessarily complex, but the jurisdiction of multiple government agencies also overlap and further slow the process. For example, a hydropower developer in Iowa must consult with 55 federal and state agencies, Native

American tribes, and nongovernmental organizations.[45] Hydropower developers throughout the nation must coordinate with the U.S. Army Corps of Engineers if they want to alter the physical structure of a dam or river. The problem of agencies overlapping is compounded by the fact that they have limited coordination, elongating the wait time for a developer's license.[46]

FERC is usually the principal agency to oversee licensing, ensure compliance, and comprehensively plan for all nonfederal dams. FERC regulates more than 1,600 hydropower projects at over 2,500 dams in the United States.[47] Any dam in the country with hydroelectric turbines must receive a license from FERC before producing electricity, unless they are federal dams built before 1920.[48] The FERC licensing process starts with deciding which type of permit process is correct for each proposed hydropower project. FERC officials can choose between the integrated, traditional, or alternative licensing processes for each hydropower project, but the integrated process is the most common.[49] In the integrated licensing process, a hydropower developer submits a proposal and information document to FERC. After the proposal is submitted, FERC seeks input from the public, nongovernmental organizations, and government agencies at the local, state, and federal levels. The developer and FERC take the concerns from the input process and form a study plan. According to FERC, studies typically take one to two years to complete.[50]

Once the studies are finished, the developer can then draft an application, which must include a detailed description of the project's facilities, the expected effects, and proposed mitigation measures. FERC officials review the application, prepare a set of environmental rules, issue environmental reports, and then seek another set of public comments on the proposed license. FERC officials then decide whether the proposed project will impair waterways, meaning that FERC must represent the interests of the federal government and the state government. After all of the considerations, comments, and paperwork are completed, FERC officials can make a final decision whether to license, modify, or decline a permit for a proposed hydropower project.[51]

In some circumstances, a whole separate licensing process with the Army Corps of Engineers becomes necessary. The Corps must become involved in the licensing process when a hydropower developer wants to construct or modify a dam, or if a developer wants to install a turbine on a non-powered dam that requires dredging. Dredging involves scooping out sediments near the dam. Installing a turbine on a non-powered dam may require dredging in some cases. Depending on what the specific hydropower project includes, a developer only needs to follow one agency's regulation, but sometimes developers have to simultaneously interact with separate agencies when their regulations overlap, adding unnecessary inefficiency. Because FERC and the Army Corps of Engineers lack comprehensive coordination, they impose their regulations at different times, which means that developers may have to fulfill identical requirements for different agencies at different times.[52]

With the combination of laws, regulations, and multiple agencies, obtaining a hydropower license often takes several years. Because there are so many moving parts in the entire licensing process, the actual costs and time frames are inconsistent and unpredictable. Lucky hydropower developers may receive a FERC license in as little as two months, but for most developers, the wait time often exceeds five years.[53]

Attempts to Fix the Regulatory Burden

It is not difficult to see that the hydropower licensing process is complicated and costly. It is also easy to see that not all hydropower is the same. Most laws and regulations do not distinguish between a Hoover Dam–size hydroelectric facility and a small turbine on a municipal water pipe. Officials in the Department of Energy and members of Congress have spoken out about the potential economic and environmental benefits of increasing the use of small-scale hydropower. Despite the lip service, over 50,000 non-powered dams and countless conduits are not electrified because prospective developers avoid the expensive regulatory barriers.

In 2013, Congress attempted to tackle this burdensome red tape by passing two laws intended to make hydropower development less costly and more straightforward. The Hydropower Regulatory Efficiency Act of 2013 and the Bureau of Reclamation Small Conduit Hydropower Development and Rural Jobs Act were designed to simplify the regulations surrounding hydropower development. In 2013, Representative Cathy McMorris Rodgers (R-WA) introduced the Hydropower Regulatory Efficiency Act (HREA) to boost hydropower in her home state of Washington, which is already heavily reliant on hydropower.[54] Washington is home to the Grand Coulee Dam, which is the largest hydroelectric power plant in the country. The state generates 58.4 percent of its electricity from a combination of large and small dams.[55] McMorris Rodgers's bill passed through both the House and Senate with little opposition. President Obama signed the HREA into law in August 2013.

The HREA exempts conduit-based hydropower plants that produce 5 megawatts or less from FERC licensing requirements, excluding those on federal lands.[56] Developers who want to build small-scale conduit-based hydropower only have to file intent with FERC. If FERC officials decide that a project qualifies for the license waiver and no formal opposition occurs within 45 days of the waiver being granted, then the project can move forward.

The HREA also allows FERC to exempt nonconduit small-scale hydropower with outputs of 10 megawatts or less. The HREA does not exempt developers from licensing requirements, but it does make the licensing process shorter and less rigorous, and exemptions do not expire.

The HREA also allows FERC to issue permit extensions for potential license applicants. Before the law was passed, permits only lasted three years, but now FERC can extend the length of the permit for two more years. These permit extensions were meant to help developers to more accurately survey sites and determine the costs of prospective projects.

Although the intentions of the HREA are noble, the legislation has done little to promote hydropower's growth. As of June 2015, only 58 small hydropower conduit-based projects have applied for the FERC licensing exemption, and from those applications, 43 were approved, 8 were rejected, and 7 were still pending. FERC received 30 applications for the two-year extension of preliminary permits and has granted 15 permit extensions, denied 14 for "lack of diligence," and 1 was still pending.[57] FERC's Robert Bell oversees conduit exemptions and claims that there was an average of 10 requests per year for conduit exemptions before the HREA.[58] The 58 exemption requests in the two years between 2013 and 2015 show an increase, but the thousands of opportunities for conduit-based hydropower suggest that the HREA may not be sufficient. It is possible that not enough time has passed for developers to start developing projects under the new regulations. In the future, the HREA may prove more beneficial than it has initially seemed.

The HREA has done little to create incentives for developing hydropower resources because it only addresses part of the licensing process for conduit-based hydropower. The HREA also does very little to combat the existing regulatory barriers to existing non-powered dams. The HREA changed which non-powered dams could qualify for traditional FERC licensing exemption. Section 3 of the HREA increases the maximum size of proposed hydropower projects that are eligible for an exemption from 5 to 10 megawatts. Although the legislation is a step in the right direction to make hydropower development simpler, it does not do anything to simplify the complex regulatory burden for non-powered dams. In essence, the HREA has not done anything to make the actual problem for non-powered dams any better.[59]

Another law that was meant to help alleviate the onerous regulations for the hydropower industry was the Bureau of Reclamation Small Conduit Hydropower Development and Rural Jobs Act. Representative Scott

Tipton (R-CO) and Senators John Barrasso (R-WY) and Jim Risch (R-ID) sponsored the legislation, which was signed into law in 2013.[60] This law was meant to facilitate small-scale hydropower in two ways. First, it authorizes power production from Bureau of Reclamation canals.[61] Second, it allows conduit-based hydropower developers to bypass NEPA's environmental assessment framework if the project's capacity is 5 megawatts or less.[62] Some people may argue that it might be dangerous to weaken the environmental assessment requirements. In this case, conduit-based hydroelectric facilities would not pose a substantial risk for two reasons. First, the risk of environmental damage is minimal because the conduits already exist. Second, canals and conduits are not usually habitats for fish, so these projects have a minimal risk of harming any fish or wildlife. By skipping the NEPA documentation, the law was meant to give developers additional time and reduced costs.

The Bureau of Reclamation has identified 373 canals that are suitable for hydroelectric power development.[63] Colorado and Wyoming have the highest potential to generate electricity from conduit hydropower on Bureau of Reclamation canals. Wyoming alone has 121 potential sites for hydroelectric generation. The abundance of Bureau of Reclamation canals in these states illuminates why Representative Tipton of Colorado and Senator Barrasso of Wyoming sponsored the bill. When pressing for the bill's passage, Senator Barrasso stressed that the bill would provide a cheap and clean source of energy and also help create jobs in rural western areas.[64]

The Bureau of Reclamation Small Conduit Hydropower Development and Rural Jobs Act is another step in the right direction to help boost hydropower, even though the law's scope is very narrow. With both the HREA and this law, legislators are beginning to realize how difficult the bureaucratic red tape has made hydropower development.

The conflicting regulations discourage investment in hydropower, despite the fact that it is one of the most economical and clean renewable energy sources. Instead of trying to make additional exemptions and small marginal policy changes, policy makers could completely reform the licensing process for small hydro projects with negligible environmental impacts. This approach along with allowing for non-electrified dams to be electrified has the potential to spur more renewable electricity generation that would not be needing the substantial subsidies so often associated with attempts to generate renewable power.

CHAPTER NINE

Nuclear

The Political Fiasco of Yucca Mountain

Since nuclear power became widespread in the United States in the 1960s and 1970s, the primary political concern has been what to do with nuclear waste. Building a long-term repository has become one of the most controversial and longest lasting political battles in the country. Yucca Mountain, located in the remote Nevada desert, was designated as the site for long-term storage of nuclear waste from nuclear power plants. Over the past three decades, the Yucca Mountain project has become one of the largest boondoggles the federal government has ever engaged in.

The failure of this government project has been slow and painful. After years of delay and billions of taxpayer dollars, Yucca Mountain is nothing more than an empty hole in the ground. Current political decisions have effectively killed the project, and it is highly unlikely the project will ever be completed. Yucca Mountain serves as a model for how politics, misinformation, and bureaucratic decision making can cost Americans billions of dollars with zero beneficial results. How did Yucca Mountain become such a massive failure? The story is a tangled jumble of back-and-forths between Congress, federal agencies, the Nevada state government, and private organizations.

Understanding Yucca Mountain's failure requires understanding the federal government's role in the nuclear power industry. The federal government oversees all nuclear power plants in the country, as well as the disposal of nuclear waste. In 1974, Congress passed the Energy Reorganization Act, which created the Nuclear Regulatory Commission (NRC) to protect public health, the environment, and national security. The NRC is the main government agency that regulates how civilians can use nuclear materials. The NRC's main function is to license nuclear facilities and regulate how the private sector uses and disposes of nuclear materials.[1]

Under the NRC's oversight, the number of nuclear power facilities grew throughout the country in the 1960s and 1970s. More nuclear power plants meant more nuclear waste to deal with. In the age of the Cold War, people were worried that nuclear waste could be turned into nuclear weapons, and nuclear waste was a top concern for lawmakers and the public alike. In 1982, Congress passed the Nuclear Waste Policy Act to deal with the growing amount of nuclear waste that was accumulating throughout the country. This law instructed the Department of Energy (DOE) to select sites and develop two long-term storage facilities for high-level nuclear waste. The law also stipulated that two repositories were supposed to be built, one on each side of the country.

The Nuclear Waste Policy Act also established the Nuclear Waste Fund that was meant to finance building this long-term repository. The money for this fund came from companies that own or operate nuclear power plants. From 1983 to 2014, the federal government imposed a fee on the electricity that these companies generated. The companies pass these fees onto ratepayers who get their electricity from nuclear power plants. Over this 30-year period, nuclear power customers have paid over $30 billion into the fund, which was meant to build the Yucca Mountain facility.[2] Since progress on the Yucca Mountain project has come to a halt, these funds have essentially been wasted, doing little to advance the projects goals of a permanent repository.

The Nuclear Waste Policy Act set up a complex interaction between several government agencies to make sure that the health of the public and the environment were overseen by multiple agencies. First, the Environmental Protection Agency (EPA) sets the standards for radiation to protect public health. The EPA's standards set restrictions on radiation exposure for people living near Yucca Mountain. Second, the Nuclear Regulatory Commission makes sure that any storage facility adequately follows the EPA's guidelines. If a storage facility like Yucca Mountain does not follow the EPA's standards, the NRC will not issue a license. Third, the Department of Energy was in charge of selecting a site for and constructing the actual facility, subject to the NRA license and the EPA's regulations.

The DOE followed the requirements of the Nuclear Waste Policy Act, and by 1983, the department had selected nine potential locations in six states. After three years of further studying these sites, President Ronald Reagan approved three potential sites: Hanford, Washington; Deaf Smith County, Texas; and Yucca Mountain, Nevada. In 1987, Congress amended the Nuclear Waste Policy Act, stating that Yucca Mountain was the only viable site to be considered for long-term waste storage. These 1987 amendments began the battle over Yucca Mountain.

For nearly a decade after Yucca Mountain was chosen as the sole site, the DOE continued to study the area's geology. By 1996, several government agencies began to realize that Yucca Mountain's geology was not as ideal as they had initially hoped. At first, the DOE had assumed that the storage facility within the mountain would require only minimal construction, using man-made caverns as the storage facility. DOE officials realized they would have to engineer complex barriers inside the mountain to meet the EPA's requirements.

In 2001, the EPA set disposal standards for the potential facility at Yucca Mountain. These standards set a maximum limit for radiation that those living closest to the facility could be exposed to. According to the standards, these limits could not be exceeded for at least 10,000 years. The EPA deemed that 10,000 years was sufficiently long enough to make sure radiation levels do not pose a significant health risk.[3]

In 2002, while the EPA, DOE, and NRC were dealing with lawsuits against Yucca Mountain, Energy Secretary Spencer Abraham officially recommended to President George W. Bush that Yucca Mountain be developed. President Bush approved the Yucca Mountain plan, but he was met with opposition from Nevada governor Kenny Guinn, who issued a Notice of Objection. This notice served as a veto to the Yucca Mountain site, but both the U.S. House of Representatives and Senate overrode the governor's objection. With this congressional resolution, Yucca Mountain was effectively designated as the official repository for the nation's nuclear waste.[4]

In 2003, the Nuclear Waste Technical Review Board released a report stating concerns with Yucca Mountain's geology. The report stated that the mountain's subterranean humidity and salt formations could corrode the nuclear fuel canisters too quickly. In 2004 the board suggested that more research was needed regarding corrosion after the first thousand years of the repository's operation.[5]

Also in 2004, the D.C. Circuit Court of Appeals ruled on the lawsuits from 2001 and found that the EPA's 10,000-year compliance period for radiation protection was illegal because it was inconsistent with research from a National Academy of Sciences report from 1995. Because the court saw that report's recommended time period was much longer than 10,000 years, the EPA was forced to start over on its rules.

By 2005, the EPA had proposed new standards. Under the Energy Policy Act of 1992, the NRC had to modify their requirements for licensing Yucca Mountain to be consistent with the EPA's new standards. For the

DOE to receive the licenses to move forward with the project, it had to wait for the EPA to finalize its new standards so that the NRC could finalize its new licensing requirements. The EPA's new standards were not finalized until 2008, which put the project on hold for a few years.[6]

During the time that the EPA was finalizing its new standards, Senator Harry Reid of Nevada became the Senate majority leader in 2006. In his position of power, Senator Reid became one of the loudest and most powerful voices against Yucca Mountain. In spite of Reid's strong and vocal opposition, the Yucca Mountain project moved slowly forward.[7]

In 2007, the DOE issued new environmental impact studies and made them available for public comment.[8] In June 2008, after the public comment period finished, the DOE issued its finalized environmental impact studies and submitted a license application to the NRC for the repository. The application showed that the DOE intended to open Yucca Mountain in 2017.[9] Shortly after the DOE submitted the application, it delayed the opening date until 2020. Also in 2008, the EPA finalized its new standards that extended the compliance requirements for Yucca Mountain from 10,000 years to one million years.[10] It was difficult to prove that the facility could protect people from radiation for 10,000 years, but it is nearly impossible to guarantee that the facility can last for a million years. This dramatic new requirement is incredibly difficult to comply with and may be unnecessary considering how little we know about the future. In March 2009, Secretary of Energy Steven Chu unexpectedly announced that the DOE would terminate the Yucca Mountain repository program. The DOE said that it would look for other options for long-term storage for nuclear waste.[11]

The DOE made its official decision to stop any work on Yucca Mountain in March 2010. In a motion to the NRC's Atomic Safety and Licensing Board, the DOE withdrew its license application "with prejudice." This type of withdrawal meant that the DOE would not consider Yucca Mountain as a repository site in the future. In June, however, the NRC's licensing board fired back, denying DOE's motion. The NRC said that the DOE would violate the Nuclear Waste Policy Act if it did not continue with the licensing process for Yucca Mountain. Legally, the NRC or the courts could force the DOE to resume the licensing process.[12]

The situation then became even more complicated. The NRC's licensing board admitted that even if the DOE were compelled to go through with the license application process, the Yucca Mountain facility may never be built. First, Congress may change the law, negating the wording in the Nuclear Waste Policy Act. Second, Congress could choose not to fund the Yucca Mountain project, which would effectively kill the program, despite any action from the NRC or the courts. Several lawsuits also emerged over the withdrawal of the DOE's license application. Some state government and private entities sued the DOE and the NRC on the grounds that the DOE could not legally terminate Yucca Mountain.[13]

When the DOE decided to stop moving forward with Yucca Mountain in 2010, it asserted that there were "better solutions" that would be less politically controversial, but did not identify what those solutions might be. Despite legal threats to the DOE, the project was effectively killed in 2011 when Congress cut federal funding for the site. Congress has not appropriated any funding for the DOE's waste disposal program or Yucca Mountain since then. Congress has appropriated less than $40 million to the NRC and other federal agencies for other nuclear waste disposal activities since 2010.[14] Shortly after the funding for Yucca Mountain was cut in 2011, the NRC stopped its licensing activities for the project. Even with the legal obligation to force the DOE to continue with Yucca Mountain, NRC leaders realized that the lack of funding meant that the project was impossible.

In 2013, however, the U.S. Court of Appeals for the District of Columbia Circuit ordered the NRC to resume its license review for Yucca Mountain. Because the NRC still had funding, even though Yucca Mountain itself did not, it finished its technical review of the DOE's application for Yucca Mountain. In January 2015, the NRC found that the DOE's license application met nearly all standards and regulatory requirements. Even with the application review complete, the NRC is still required to resolve challenges from parties with legal standing in the licensing process, including Nevada, Indian tribes, and private industries. Although Yucca Mountain has no funding, the NRC will still take several more years and spend taxpayer dollars to resolve the challenges brought against a dead government project.[15]

In 2011, the Government Accountability Office (GAO) investigated the reasons why the DOE decided to terminate the Yucca Mountain project. The GAO found that the real death blow to Yucca Mountain was "social and political opposition to a permanent repository, not technical issues." The DOE stated that there were "better solutions" for nuclear waste than Yucca Mountain, but it failed to identify what any of those alternative solutions would be. The DOE officially responded to the NRC's licensing board, stating, "the Secretary's judgment is not that Yucca Mountain is unsafe or that there are flaws in the license application, but rather that it is not a workable option and that alternatives will better serve the public interest."[16]

In an interview in December 2015, Senator Reid gloated about the continued lack of funding for the project, saying, "Notice there is nothing in these bills for Yucca Mountain." Reid also decreed the Yucca Mountain project was "just a big boondoggle," conveniently omitting the fact that it was mainly his political tactics that had turned the multi-billion-dollar operation into a boondoggle.[17] Senator Reid's actions are easily explainable. As a self-interested politician, he worked to do what he thought his constituents wanted. Nobody really wants to live next to the country's nuclear waste dump. Local opposition to the project represents a classic case of the "not in my backyard" mindset.

Nuclear power companies and other nongovernmental organizations who have nuclear waste now have no option for long-term storage. High-level nuclear waste is stored on-site at various nuclear facilities scattered across the country. Over 70,000 metric tons of nuclear waste from commercial nuclear power plants is spread across dozens of locations in the country. Each year, the amount of nuclear waste from nuclear power plants grows by roughly 2,200 metric tons per year. Just to give some perspective, if all nuclear fuel waste was piled on a football field, the pile would be over 55 feet tall.[18] Seven miles of tunnels have already been dug under Yucca Mountain in preparation to store the nation's nuclear waste, but those tunnels will likely sit empty unless major political decisions are reversed.[19] With the political death of Yucca Mountain, the United States has no permanent repository for nuclear waste, and the future of nuclear power is unclear.

Nuclear Power and Government Oversight

Nuclear power has been in use since the 1950s. Many of the policies and public sentiments regarding nuclear power reflect the fears and prejudices of the Cold War era. When many people hear the term "nuclear power," their first thoughts go to Chernobyl, Three Mile Island, or Fukushima. These disasters are important and should not be overlooked, but many countries have used nuclear power every day for decades without major problems. Nuclear power is one of the largest sources of electricity in the United States, providing about 20 percent of the country's electricity.[20]

France is the most nuclear powered country in the world, producing 76.3 percent of its electricity from nuclear power. Ukraine, Slovakia, and Hungary follow France as the other most nuclear countries, each producing over half of their electricity from nuclear power.[21] Even though some countries may produce a larger percentage of their electricity from nuclear power, in terms of output, the United States is by far the largest producer of nuclear-generated electricity. In 2015, American nuclear power plants generated 771 billion kilowatt-hours, which is enough electricity to power approximately 70 million average American homes.[22]

Nuclear power is also one of the most efficient types of electricity generation and has one of the highest capacity factors of any major energy source, with an average of 92 percent. Nuclear power plants generate a higher percentage of their capacity than other sources, running almost constantly because they are difficult to ramp up and down. One main advantage of nuclear power is it provides large amounts of consistent baseload power.[23]

In the United States, 60 commercial nuclear power plants in 30 states provide the country with one-fifth of its electricity. The largest nuclear power plant in the country is the Palo Verde power plant, located in Arizona. As of 2014, Palo Verde has the capacity to generate 3,937 megawatts of electricity, which is roughly equivalent to two Hoover Dams operating at full capacity.[24] The smallest nuclear power plant is the Fort Calhoun Nuclear Generating Station in Nebraska, with the capacity to generate 479 megawatts.[25] The waste from these commercial nuclear power plants is stored at 75 reactor sites in 33 states across the country.[26]

Since 1974, the NRC has regulated commercial nuclear power plants. Although there are several types of nuclear reactors, the United States only has pressurized water reactors and boiling water reactors that produce electricity commercially. In the simplest terms, nuclear power plants use the heat from radioactive materials to turn water into steam, which is used to power an electricity-generating turbine. Of the 100 licensed nuclear power plants in the United States, 66 are pressurized water reactors and 34 are boiling water reactors.[27]

Pressurized water reactors work by using the heat from the nuclear "core" to heat pressurized water in the primary coolant loop. The primary coolant loop leads to the steam generator, where the heat from the primary coolant loop is transferred to water in a secondary loop, vaporizing into steam. The steam from the secondary loop goes to the main turbine, spinning the electric generator, which produces electricity. The steam is condensed back into water, sent back to the steam generator to be reheated, and then used to power the turbine continuously.[28]

Boiling water reactors use the core to heat water into a steam-water mixture. The steam-water mixture exits the top of the core and enter separators to remove water droplets from the steam. Once water droplets are removed, the steam goes directly into the main turbine, which spins the electric generators. After the steam exits the turbine area, it is sent to a condenser and then pumped back to the reactor to begin the cycle over again.[29]

The number of power plants in the United States is not changing. In the past few years, a total of four nuclear power plants in Florida, Wisconsin, Vermont, and California were taken off-line. The Watts Bar Nuclear

Generating Station in Tennessee added a new reactor, which is expected to be commercially operational by the end of 2016. New nuclear reactors are being added to existing nuclear power plants. In Georgia, two new reactors are being added to the Vogtle Electric Generating Plant, also known as Plant Vogtle. The Virgil C. Summer Nuclear Generating Station in South Carolina is also adding two new units.[30]

Expansion of the nuclear industry has stagnated for a few decades, thanks to unfavorable policies, political quagmires, and misinformation. In its early years, nuclear power was not as controversial as it is today. President Dwight D. Eisenhower was a strong advocate for nuclear power in the 1950s. On December 8, 1953, Eisenhower delivered his "Atoms for Peace" speech to the United Nations, propelling the idea of nuclear power into the national and worldwide consciousness. Eisenhower worked with Lewis Strauss as the chairman of the Atomic Energy Commission (AEC) to encourage utilities and private companies to construct nuclear power plants. In February 1954, the AEC began a reactor-development program that used government funding to spur innovation on nuclear reactors. Eventually the light water reactor was the most successful.[31]

Congress also began helping the nuclear industry with the Atomic Energy Act of 1954. Under this act, private companies could own nuclear power plants. The law set up a system where private companies would work with government agencies to produce large amounts of electricity in a way that would not harm the health of humans or the environment while also protecting national security. Private companies would pay for construction and plant operations, and the government would perform research and development, offer technical assistance, and be responsible for disposal of nuclear waste.[32]

With support from the executive and legislative branches, the nuclear industry in the United States began to grow. Shippingport Atomic Power Station became the United States' first commercial power plant, opening on May 26, 1958. The power plant was built on the Ohio River west of Pittsburgh, Pennsylvania, at a cost of $72.5 million.[33]

Following Shippingport, many companies constructed nuclear power plants throughout the country, but Cold War concerns dominated the policy field. One of the biggest concerns was what to do with nuclear waste after it was used to generate electricity. Dealing with nuclear waste is much more of a political problem than a technological one. Many countries such as France, Russia, China, and Japan recycle and reprocess their nuclear waste and generate energy from it. Once nuclear fuel is used once, it still contains energy potential. Spent nuclear fuel can be reprocessed and used to generate energy a second time.[34]

Recycling nuclear waste can be beneficial in some circumstances. First, recycling nuclear waste reduces the amount of high-level waste to roughly one-fifth of what it would have been. Second, recycled waste is less radioactive and does not need to be stored as long until it is no longer hazardous. There are currently no reprocessing plants for nuclear waste in the United States. In the 1960s and 1970s, however, three reprocessing plants were built. The last of these plants closed in 1977 when U.S. federal policy outlawed all civilian reprocessing in its nonproliferation policy.[35] President Jimmy Carter helped further entrench the misperception that waste reprocessing was dangerous through his administration's nuclear policies. Burying nuclear waste became the only viable option, and this myopic view still haunts the nuclear industry today.[36]

Reprocessing nuclear waste is still controversial. Some people argue that reprocessing would increase the risk of nuclear terrorism, increase the ease of nuclear proliferation, and hurt U.S. nuclear waste management efforts. The fears of nuclear proliferation or terrorism are the same today as they were in the Cold War. If nuclear materials were to get in the wrong hands, they could be dangerous, but reprocessing facilities are not the only way that this can happen. Perhaps the strongest argument against reprocessing is that it would take focus and finances from the current waste management effort. Considering that Yucca Mountain is politically unviable for the foreseeable future, long-term waste management is already a major problem. On the other side of the argument, supporters of reprocessing think it would reduce the need for nuclear waste storage and disposal. Reprocessing would help reduce the amount of nuclear waste that needs to be stored, but it would not completely eliminate the need for storage and disposal.[37]

Those who argue that recycling and reprocessing nuclear fuel is not a viable solution should look to the rest of the world. Many countries reuse nuclear fuel without substantial safety or environmental issues. The big question in the United States comes down to whether reprocessing would be economical. Policy makers will have to consider the relative costs of reprocessing versus not reprocessing now and into the future. Some researchers argue that reprocessing costs and direct disposal are nearly equal. Others assert that reprocessing will be more cost-effective as uranium prices increase, and other analyses find that the margin between the cost of reprocessing is much higher than disposal and will continue to be for several decades.[38] More research is needed to know how cost-effective reprocessing could be in the United States, but we will not know exactly how cost-effective it could be unless entrepreneurs and innovators are able to work with nuclear waste reprocessing. However, unless current regulations are reformed or removed, this innovation is unlikely to occur.

Fees and Licenses for a Nuclear Power Plant

The Office of Nuclear Reactor Regulation (NRR) is a sub-branch of the Nuclear Regulatory Commission and is responsible for regulating all commercial nuclear power plants in the nation. In particular, the NRR engages in four regulatory actions: rulemaking, licensing, oversight, and incident response. The NRR has several subdivisions that oversee specific areas of the regulatory process. Some of the most important divisions are the Division of Policy and Rulemaking, the Division of Operating Reactor Licensing, and Division of Inspection and Regional Support.[39]

The NRC's licensing process can give a company the authority to build, operate, or decommission commercial reactors. Any company that wants to engage in the nuclear power business must receive a license through the NRC. The NRC also needs to be involved if an existing license is amended, renewed, or transferred. When NRC staff reviews an application for a license, their primary goals are to "ensure that the applicant's assumptions are technically correct" and that "the proposed activities will not adversely affect the environment."[40]

The NRC collects fees from the companies that it oversees and licenses, including commercial nuclear reactors, facilities that make nuclear fuel, uranium mining companies, and hospitals that use machines with nuclear material. Under the Independent Offices Appropriation Act of 1952 and the Omnibus Reconciliation Act of 1990, the NRC must recover 90 percent of its annual budget from the companies it licenses and oversees. Each year, Congress sets the NRC's budget, and when the budget changes or the number of license applicants changes, the NRC adjusts its fees so that the fees recover 90 percent of the allocated budget. The agency does not use the fees directly. All income from fees is sent to the General Treasury.[41]

The NRC charges annual fees to recover generic regulatory costs and also hourly fees to recoup the costs that NRC staff incur reviewing each company. For fiscal year 2015, Congress appropriated slightly over $1.015 billion to the NRC, meaning that the agency is required to recover $895.5 million through annual and hourly fees. Hypothetically, a licensed company may pay $6 million in annual fees, and the same company could pay an additional $200,000 in hourly fees if it requested an amendment to its license. For fiscal year 2015, the NRC amended its flat application fees to an hourly rate of $268.

NRC fees are subject to change each year, and come in many forms that collectively add up to hundreds of millions of dollars. For fiscal year 2015, the NRC had to recover $567 million through annual fees from current licensees, so it adjusted its annual fees to recoup that cost. For example, all operating power reactors were charged an annual fee of $4.8 million with an additional $223,000 for spent fuel storage or reactor decommissioning, totaling a combined fee of $5.03 million. High enriched uranium fuel facilities were charged an annual fee of $8.5 million and low enriched uranium fuel facilities were charged $2.9 million in fiscal year 2015. Dozens of other annual fees apply to other companies who are licensed through the NRC.[42]

Fees are not the only costs that nuclear power companies incur. The time and monetary costs for following the many steps in the licensing process can deter would-be nuclear power producers. Some people may argue the lengthy licensing process and all its costs are necessary for public safety and environmental health. Regardless of how necessary or useful the current licensing process is, it is undeniably complex, lengthy, and costly.

To obtain an NRC license for a nuclear power plant, companies must receive a construction permit and an operating license. For a construction permit, the NRC reviews the applicant's preliminary plant design and the suitability of the prospective site. On the application for a construction permit, a company must include preliminary safety analyses, an environmental review, and financial and antitrust statements. The applicant also has to include an assessment of market demand for the power plant. If NRC staff members decide that the construction permit application meets all the requirements, then the agency publishes a notice of receipt in the Federal Register. After publishing the notice of receipt, the NRC reviews the application in a safety evaluation report.[43]

Public involvement is a large part of the construction permit. The NRC sets up public meetings near the proposed site so the public can have basic knowledge about the project, including safety, environmental impacts, the planned location, the type of plant, and the licensing process. Over the course of the permit process, the NRC and the applicant holds other public meetings on the plant's design and construction. The Advisory Committee on Reactor Safeguards serves as an independent group of technical experts to advise and review each construction permit application. The advisory committee prepares a report of its reviews to the NRC.[44]

The NRC also follows the National Environmental Policy Act to evaluate the potential environmental impacts and benefits of the proposed plant. The NRC issues a draft environmental impact statement, which the public and agencies at the federal, state, and local levels can comment on. After the comment process, the NRC revises the environmental impact statement, and issues it in its final form.[45]

Once the final environmental impact statement is issued, the applicant submits final plans for operation and an application for an operating license. This application requires a final safety analysis report and an updated environmental report, which includes the final design, a safety evaluation, operational limits, and plans for emergencies. The NRC consults with the Federal Emergency Management Agency to review the emergency plans to make sure they are adequate. After the Federal Emergency Management Agency gives its approval, the advisory committee holds a public meeting where it reviews both the operating license application and the NRC's final safety evaluation report. After the advisory committee's review, the NRC published a notice in the Federal Register for anyone whose interests might be affected. The Atomic Safety and Licensing Board holds the public hearing, and makes its decision with the same process as for the construction permit hearing. After passing all these steps and conferring with all the different agencies, a nuclear power developer can receive a license.[46]

There are several different processes a developer can use to meet all the requirements for a nuclear operating license, but the most common is the combined license (COL). A combined license is used to authorize both construction and conditional operation of a nuclear facility, combining multiple licensing processes into one. According to the NRC, this process typically takes anywhere from 44 to 74 months.[47] With enough public opposition, however, the licensing process can be extended to several decades.[48]

Nuclear Waste Issues

The Nuclear Waste Policy Act of 1982 (NWPA) was meant to help solve the nuclear waste problem by providing a way to dispose of commercial waste. The NWPA not only started the process of underground storage at Yucca Mountain, but it also gave the DOE the responsibility to remove waste that was stored at individual power plants and transport that waste to disposal facilities owned by the federal government. The DOE signed contracts with commercial nuclear power producers to start removing waste by 1998. The DOE did not meet its deadline in 1998, so electric utilities began to sue for damages. As of the end of 2015, 35 of these lawsuits have been settled, and another 33 cases have been resolved by final judgments, with 19 cases still pending. The federal government has paid these reparations from the Treasury's Judgment Fund, which allocates taxpayer money to pay for lawsuits filed against the government. We can expect commercial nuclear power plants to file suits in the future as they try to recover the costs from storing nuclear waste longer than they had contractually agreed.[49]

More than 17 years have passed since the DOE was supposed to begin collecting commercial nuclear waste in 1998. Not only has the political death of Yucca Mountain been a waste of taxpayer dollars, it will continue as long as the DOE does not have a place to store the waste that it is legally required to dispose of. By breaching its contracts with electric utilities and losing in lawsuits, the federal government has paid $5.3 billion in damages to electric utilities from the Judgment Fund. The reimbursement to each nuclear power company is unique based on several considerations, such as the age, operating status, and amount of available waste storage.[50]

It is easy to assume that if the DOE cannot fulfill its obligations to transport and dispose of commercial nuclear waste in the near future, taxpayers will continue to bear the burden of a department that must pay utilities for a service that it is incapable of providing. We can expect that the Treasury's Judgment Fund will continue to pay out billions of taxpayer dollars because political infighting has blocked the ability of the DOE to fulfill its legal obligations. If the DOE is able to start accepting nuclear waste within the next 10 years, it estimates that remaining liabilities will total an additional $23.7 billion.[51]

In 2013, the U.S. Court of Appeals for the DC Circuit ruled that the DOE could no longer collect fees from nuclear power companies for the Nuclear Waste Fund. The court ordered the secretary of the DOE "to submit to Congress a proposal to change the fee to zero until such time as either the Secretary chooses to comply with the Nuclear Waste Policy Act as it is currently written, or until Congress enacts an alternative waste management plan." The court also said, "Until the DOE comes to some conclusion as to how nuclear wastes are to be deposited permanently, it seems quite unfair to force petitioners to pay fees for a hypothetical option, the costs of which might well—the government apparently has no idea—be already covered."[52]

Since the Nuclear Waste Policy Act was passed in 1983, the DOE has spent roughly $15 billion just to study potential storage sites, research the Yucca Mountain site, and develop the license application. About $9.5 billion from that $15 billion was from the Nuclear Waste Fund.[53]

The Government Accountability Office (GAO) published a report that gives two broad suggestions for America's nuclear waste management in the future. The GAO first suggests that the most important obstacle to a permanent repository for nuclear waste is not a technical or safety issue. The biggest problem is social and political opposition, which means that politicians and bureaucrats will need to consider increased transparency, economic incentives, and education if they are going to find a sustainable solution. Second, the GAO suggests that consistency is a key part of a successful waste management strategy. The process of selecting, licensing, and constructing a permanent repository will take decades, as we saw with

Yucca Mountain, and without consistency in funding and how policies are implemented, billions more dollars could be wasted.[54]

One potential solution for nuclear waste management that the GAO did not look at is a more market-oriented approach. Much like commercial nuclear power plants can make profits, nuclear waste disposal may also be able to be accomplished commercially. Some argue that nuclear waste is much too dangerous to be allowed to be done by profit-driven companies. This logic is not consistent with the current regulatory regime especially considering that commercial nuclear power plants are also potentially dangerous, yet we still have many of them throughout the country. Deciding exactly how a nuclear waste market should be set up and which regulations it should be subject to is a complicated question. Our purpose here is not to suggest a particular type of nuclear waste market, but we do suggest that policy makers should consider the power of markets and market incentives to accomplish goals. One thing is certain—the status quo under complete government control of nuclear waste has been a decades-long disaster with no real results other than billions of wasted taxpayer dollars.

If the DOE had finished building Yucca Mountain, the United States would likely have had a permanent solution to nuclear waste and less uncertainty over the future of waste-related safety. The Obama administration, key congressional leaders, and the state of Nevada became an insurmountable blockade to the project. Amid all the uncertainty surrounding the future of nuclear waste policy, it seems certain that politics will continue to play a role in how the United States resolves its issues with nuclear-generated electricity.

CHAPTER TEN

Coal

Clean Air, Poor People in the Southwest

Near the red cliffs and blue water of Lake Powell, three smokestacks loom over the desert landscape, visible for dozens of miles in every direction. This coal-fired power plant, known as the Navajo Generating Station, is the largest in the West and provides electricity for several southwestern states. The power plant sits on the red sands of the Navajo Nation near the Arizona-Utah border. Although the Navajo Nation struggles with widespread poverty, the Navajo Generating Station is also one of the largest employers on the reservation. The power plant employs several hundred people directly, most of them Navajo. Hundreds more Navajo people work in mining operations that supply coal to the power plant, and others work on the railroads that ship coal from the mines to the plant. Although coal power in general has become increasingly controversial in the past several years, few power plants are as controversial as this one.

Because of its picturesque location and its proximity to national parks, monuments, recreation areas, and wilderness areas, many environmentalists and recreationists want to see the Navajo Generating Station closed down and dismantled. Millions of people from around the world travel to Glen Canyon National Recreation Area and Grand Canyon National Park each year. Some people fear that these natural wonders may be harmed by emissions from nearby power plants. The coal industry, however, is one of the few profitable economic ventures on the otherwise impoverished Navajo Nation, which makes the fight over what to do with the power plant even more intense.

The Navajo Nation is much like a third-world island inside the richest country on Earth. The reservation's unemployment rate is 50 percent, and a similar percentage of the population lives below the national poverty line. Nearly a quarter of Navajo families make less than $10,000 a year in household income.[1] Even though the Navajo Nation is the largest reservation in the United States in both square mileage and population, a vicious cycle of poverty keeps tens of thousands of Navajo people well below the standards of living of other American citizens.

The Navajo Generating Station's lease will expire in December 2019, and power plant officials are working with the government of the Navajo Nation to renew the lease before it expires. The fact that the plant is located on tribal land makes the lease renewal especially complicated. Tribal land is overseen by a complex bureaucracy of overlapping levels of government including the tribal government as well as federal agencies like the Bureau of Reclamation, the Bureau of Indian Affairs, and the Environmental Protection Agency.

In 2013, the Navajo Nation Council approved a measure to extend the lease for the power plant through 2044, but the lease amendments include environmental review under the National Environmental Policy Act and the Endangered Species Act.[2] Even if all goes well with the lease renewal, the EPA's regional haze regulations make the continued operation of the power plant difficult. These regulations are supposed to improve visibility for the national parks and recreation areas around the power plant, but they also make it more difficult for plants like the Navajo Generating Station to stay in operation.

One of the loudest voices against the EPA's regional haze regulations is Arizona state senator Carlyle Begay. Begay, who is of Navajo heritage, recently switched from the Democratic Party to the Republican Party largely because the EPA's regulations, which most Democrats support, will economically harm some of the poorest people in the country. Begay supports the Navajo Generating Station and the coal industry on the Navajo Nation. He said,

> Historically power plants and mines have been some of the best jobs for those in the nation, but the plants do more than just provide high paying jobs—they have allowed the Navajo Nation to preserve its unique cultural identity. . . . If you look at the EPA's stated goals and then look at how they are applied in practice, the new regulations will have a disproportionately negative impact on the economy of Navajo Nation to much greater extent than other communities in the Southwest. The EPA needs to understand that the Navajo Nation has a job crisis and a 52 percent unemployment rate.[3]

This conflict between environmental and economic goals clearly demonstrates the trade-offs inherent in any policy decision. In this case, the trade-off is one between cleaner less hazy air and the livelihoods of a severely impoverished people. As with all political questions, the answer to which priority is more important depends on who you ask. When weighing the trade-offs between more or less stringent regulations, we must ask ourselves, "How much cleaner will the air actually be, and how badly will people's livelihoods be affected if the plant were to shut down?" With the EPA's regulations looming over the future of the Navajo Generating Station, many people are trying to find accurate answers to this question as they try to balance environmental concerns with economic ones.

As part of the regional haze rule, the EPA issued a new regulation in 2014 to reduce the amount of nitrogen oxides emitted from coal power plants. The regulation, called the best available retrofit technology (BART), would require the Navajo Generating Station to install selective catalytic reduction (SCR) technology to reduce these emissions. For the technology to work, the Navajo Generating Station would also have to install additional filtration technology. The combined cost of complying with these regulations could be as high as one billion dollars, and would also lead to higher electricity rates.[4]

Even if the Navajo Generating Station installs the new equipment, it is uncertain how much better the visibility will be in the Grand Canyon and the surrounding area, but what is certain is that thousands of people will have to pay higher electricity rates to compensate for costs of the technology. If the power plant cannot afford to comply with the regulations, and instead is forced to shut down, several thousand people who directly and indirectly supply coal to the power plant or who work there will lose their jobs.[5]

Haze regulations are just one example of a whole suite of new requirements and restrictions currently being planned or implemented. Carbon dioxide emissions, for example, may become more strictly regulated in the near future, raising costs even further and imperiling even more jobs. Through all of the haze of this regulatory uncertainty, only one thing is clear: the trade-offs between environmental and economic concerns will only become more pronounced in the coming decades.

King Coal

For decades, the United States produced more of its electricity from coal than from any other source of energy. In 1980, 50 percent of America's electricity came from coal.[6] By 2013, about 39 percent came from coal.[7] Just two years later, coal was producing only 33 percent of the nation's electricity, tying with natural gas for the top spot.[8] Although production of electricity from coal is declining, the United States still has more recoverable coal reserves than any other country on Earth with enough to last for over 250 years at our current consumption rates.[9]

Coal is used to generate electricity in two ways. First is conventional coal pulverization, which involves grinding coal into a powder and burning it to boil water into steam. The steam then powers electricity-generating turbines.[10] Second is integrated gasification combined cycle (IGCC), a much more complicated process where coal and other materials are converted into a gas and ignited to boil water for the steam-powered turbines.[11] Using either of these systems can generate reliable baseload electricity.

Although we still get more of our electricity from coal than from any other source, coal may be the most controversial electricity source in use today. One on side, coal proponents assert that coal gives us a cheap energy source that provides us with reliable electricity. Cheap and reliable electricity, they argue, is a key component for living prosperous lives and promoting human well-being. The U.S. Energy Information Administration estimates that the country can continue extracting coal at current rates for 250 years before the recoverable reserves are exhausted.[12] On the other side, coal opponents argue that coal destroys the environment by creating harmful emissions that exacerbate climate change and harm human health. Both of these opinions are correct. Coal is a cheap, abundant, and reliable energy source that has enabled American prosperity. Coal also has undeniable environmental impacts, which is why many government officials are trying to discourage its use.

King Coal is falling fast because of a combination of federal and state policies that discourage coal-generated electricity in favor of generation from renewables and increased competition from natural gas. State mandates for renewable energy sources have decreased coal's market share by requiring that a minimum percentage of a state's electricity come from sources designated as renewable, while subsidies for other energy sources distort energy markets away from coal.

Regulations are costly, no matter what level of government they come from because they prevent individuals and companies from acting as they otherwise would, usually leading to higher costs for consumers. Many people support regulations, especially those with good intentions like protecting the environment. But no one can deny that some may lose when something is regulated, or that others may lose when something is not regulated. The main reason that coal is regulated is due to fears about its environmental impacts. To reduce the negative environmental impacts of burning coal, regulations often require coal power plants to install "scrubbers" to filter pollution and carbon capture technology.[13] These technologies can help reduce the carbon and pollution emissions from a coal plant, but they do not completely eliminate them, and the expensive technology also leads to higher costs for coal-generated electricity.

On top of government policies, coal is competing in the market against natural gas.[14] The shale boom of the early 2000s came about because of the widespread use of fracking and horizontal drilling, which made natural gas much easier and cheaper to produce. The lower prices of natural gas have gradually chipped away at the coal industry's large share of the electricity market.

Because of the controversy over coal's environmental impacts, the coal industry faces a tumultuous political climate. Politicians from fossil fuel–rich states in both the east and the west push to keep unfavorable policies from damaging the coal industry that is the livelihood for many of their constituents. Over the past century, much of the coal production in the United States has moved westward. Coal mining was once centered in Appalachia, but policies like the Clean Air Act have helped to shift coal mining to the mountains and deserts of the West.

In states where the coal industry is small or nonexistent, environmental sentiments have pushed those politicians to crack down on the coal industry as a dirty and archaic energy source. Special interest groups are aiding these politicians to smother the coal industry. As with many regulations, those who reap the rewards but do not pay the costs of regulation are the most eager to impose it. For example, the Beyond Coal campaign is a push by the Sierra Club and Michael Bloomberg in recent years to kill coal plants. The campaign claims that it has helped to close or repurpose 188 coal power plants across the country, while also advocating for renewables to replace coal.[15,16] The reality of New York's electrical supply provides an illustration of just how former New York City mayor Bloomberg could advocate policies that increase electricity costs. A relatively small amount of New York's electricity is produced from coal-fired power plants and the same is often true of the most ardent and concentrated environmental supporters advocating the end of the use of coal to generate electricity. They face few of the direct costs of these policies.

Discouraging or encouraging the use of coal for electricity production comes with trade-offs that politicians, bureaucrats, and special interest groups have been fighting over for many years. The many competing interests and values surrounding coal have caused federal and state policies to grow into a strange, convoluted mess. When the coal industry and environmentalists lobby for opposing goals, the resulting policies can simultaneously make coal harder and easier to use as an electricity source.

Contradictory Federal Policies

Federal policies that affect the coal industry are contradictory, because they both encourage and discourage the production of electricity from coal. Government policies that discourage coal interfere with energy markets and make it more costly to generate electricity. When coal companies and utility providers incur greater costs, their only option is to pass those costs on to electricity consumers. Government policies that encourage coal production, like tax breaks and subsidies, distort the electricity market on the taxpayers' dime. Although subsidies appear to reduce cost, in reality they simply transfer that cost to taxpayers who are forced to fund these subsidies.

Some costs from coal-generated electricity are much easier to see than others. The obvious costs include power plant construction, operation and maintenance, and transmission infrastructure. Any power plant of any kind has these costs, and they are factored into the electricity rates that utility companies charge their customers.

Other costs are much less obvious. These costs include the indirect costs that government policies and regulations create. We pay direct costs through our monthly electricity bill, but we pay indirect ones through higher taxes. Government subsidies, mandates, and regulations distort electricity markets because that is precisely what they are designed to do. These government interventions will inevitably have an impact on both taxpayers and consumers.

Some federal policies subsidize coal to make it more economical as an energy source. In 2013, coal received 6 percent of federal subsidies for all electricity sources.[17] Other energy sources, such as wind and solar, receive much larger percentages of federal subsidies.[18]

Policy makers justify intervening in the energy market because they say they want to alleviate social and environmental costs. In economics, these costs are called negative externalities, which occur when one person's actions unintentionally place a cost or a burden on someone else. Many policy makers try to overcome these external costs, or externalities, by regulating one action or subsidizing an action that they think has fewer impacts.

For coal, the social and environmental costs can include potential health problems from pollution and the environmental harm caused by mining and emissions. One of the biggest environmental concerns with coal is that burning coal leads to carbon emissions, and many are concerned that increased carbon emissions will exacerbate climate change. On the other hand, coal has the social benefit of providing cheap, reliable electricity that

Americans have depended on for many decades. The tension between the costs and benefits has made coal one of the most controversial electricity sources.

In theory, policy makers could more effectively form policies to address carbon emissions if they had an accurate way to price the effects of emitting more carbon. In recent years, many researchers and analysts have tried to put a price on carbon emissions, which would make it easier to quantify the effects of burning coal. Other people argue that a carbon tax would be a good way for coal power plants to pay for their negative impacts. The problem with assigning a price to carbon or passing a carbon tax is that there is not a concrete way to put a dollar amount on carbon. Depending on who you ask, the "proper" carbon tax can be anywhere from $5 to $100 per ton. The estimates for carbon dioxide–caused damage range between $5 to $35 per ton.[19] Because the effects of carbon emissions will not be seen until sometime in the future, there is no way to know exactly how much damage will occur. This makes it very difficult to put an accurate price on carbon emissions, even with the best theoretical backing and models.

Assigning a price to carbon is even more difficult because it is subject to the messy policy-making process. When the government tries to assign prices to something outside of a market or tries to pass a tax, special interest groups get involved for their own benefit. People in the coal industry will lobby for a lower tax on carbon, and environmentalists will lobby for a higher tax. Political games, back room deals, and favor trading become the formula for ascribing a price or a tax, instead of science.

Estimating costs is more of an art than a science because of uncertainty and subjective preferences. Markets tell us the price of something because sellers charge a price that people are willing to pay. For carbon, there is not a true market because regular people do not buy and sell carbon dioxide emissions. Without a functioning market, any price calculation will vary. Researchers and analysts will always come up with different numbers depending on the factors they include and their personal values. Economist Robert Pindyck from the Massachusetts Institute of Technology notes that models that attempt to quantify the social cost of carbon have "crucial flaws that make them close to useless as tools for policy analysis."[20]

Although assigning a specific number to the cost of carbon is difficult, if not impossible, the cost still exists. We simply do not know what that cost is. The value of reducing carbon emissions is also subjective, and highly dependent on income. People with different income levels across the world may be willing to pay very different amounts to reduce carbon emissions. A high-income individual living in the United States may say that she is willing to pay more for her electricity if she knows it is coming from renewables and helping to reduce carbon emissions. But for someone who is living in poverty and working just to get by, it may not be worth it to them to pay more to reduce carbon emissions.

Over the last 50 years, the federal government has passed environmental and health laws to combat the negative impacts from coal power plants. Two of the most pressing issues have been the emissions of sulfur dioxide and nitrogen oxides.[21] Starting with the Clean Air Act of 1970, the EPA has regulated these emissions. Over the past several decades, these regulations have become increasingly stringent. As regulations have become more stringent, the cost of compliance for coal power plants has increased. These costs get passed along to energy consumers. Of course, most regulations certainly come with some benefits. One of the most important aspects of analyzing regulations is figuring out whether costs outweigh the benefits. Another important part of analyzing regulations is understanding the unintended consequences that accompany the positive aspects of any regulation. Some of the consequences may include shifting the costs of a regulation, and when costs are shifted to consumers, regulations can end up punishing people they were not meant to punish.

The EPA also regulates other emissions, including mercury. Coal-fired power plants are the largest source of mercury emissions in the United States, releasing about 49 percent of all annual emissions.[22] When human beings are exposed to mercury at certain levels, this exposure can result in neurological and developmental disabilities.[23] In 2012, the EPA began regulating mercury emissions from fossil fuels under the Mercury and Air Toxics Standards, also known as the "Mercury Rule." The Mercury Rule was controversial since it was first proposed. The EPA asserted that if the rule were fully implemented, it would "prevent 90 percent of the mercury in coal burned in power plants from being emitted to the air."[24] The EPA also estimated that the Mercury Rule would provide from $37 to $90 billion of benefits to the American people.[25]

After the EPA issued the Mercury Rule, 21 states filed suit, claiming that the agency did not consider the costs of the regulation before deciding to regulate.[26] The Clean Air Act (CAA) spells out certain emissions that the EPA was required to regulate, but mercury is not specified in the CAA. The EPA can regulate nonspecified emissions only when "necessary and appropriate." Although mercury is not specified in the CAA, EPA officials deemed it necessary and appropriate to regulate. The plaintiffs in the suit argued that accounting for costs is a necessary step in determining if a regulation is necessary and appropriate. The EPA's counterargument was that the CAA forbids the EPA from taking costs into account when regulating emissions for sulfur and nitrogen, so the same rule would apply to mercury regulations. In June 2015, the Supreme Court favored the plaintiffs' argument and struck down the Mercury Rule. The Supreme Court also instructed lower courts to determine how to calculate cost of mercury regulations.[27]

Even though the Mercury rule was short-lived, it had a lasting impact on the coal industry. Many coal producers had already paid the costs for mercury compliance before the regulation went into effect because they had to comply with it by April 2015.[28] Because the rule was so costly to comply with, 58 coal power plants around the country had to either reduce their output to below capacity or shut down completely.[29] The Mercury Rule was a driving factor that took about 16,000 MW of capacity off the grid. That is enough capacity to power roughly 16 million American homes.[30]

Federal regulations add costs to coal-generated electricity, but the federal government also favors the coal industry with subsidies, tax expenditures, and other privileges. Special treatment for the coal industry has persisted for many decades. Some estimates state that the federal government has awarded the coal industry $72 billion (in constant 2007 dollars) from 1950 to 2008. Of that $72 billion, $31 billion has gone toward research and development, $27 billion has gone toward tax expenditures, $11 billion was paid out through the Alternative Fuel Production Credit, and $2.9 billion has been awarded through clean coal technology research and development.[31] Tax expenditures provide the largest amount of federal support for coal-generated electricity. In fiscal year 2013, tax expenditures for coal amounted to $779 million.[32]

Tax expenditures work in several different ways and are subdivided into two main categories: tax deductions or tax credits. Deductions and credits work slightly differently, but both of them artificially lower production costs. Tax deductions consist of special exemptions or accounting measures that reduce how much a company pays in taxes, as long as the company meets the government's specific standards. Through deductions, companies can keep a larger chunk of their profits, which ultimately serves as a subsidy.[33] For example, some tax deductions reduce the amount that companies pay in taxes by factoring in the depreciation of equipment.

Tax credits can be either "nonrefundable" or "refundable." If a company is eligible for a nonrefundable tax credit, the tax credit could potentially reduce the amount the company owes in taxes to zero. A company cannot make money off a nonrefundable tax credit, but they can potentially cover all the taxes they owe. Refundable tax credits, however, can actually result in cash payments if a company's tax liability is reduced to below zero. If a company is eligible for a refundable tax credit, it can eliminate any taxes it owes, and then it can actually make money if it qualifies for more money under the tax credit than it owes. In effect, refundable tax credits become a direct subsidy when corporations qualify for more tax credits than they owe in taxes, leaving taxpayers holding the bill.[34]

For example, tax credits for clean coal apply to coal producers who invest in government-approved clean coal technology. As of 2009, $3 billion in these type of tax credits was handed out to the coal industry.[35] The Energy Policy Act of 2005 (EPAct) was the first time Congress passed a tax credit for clean coal. Under this act, the federal government appropriated $1.65 billion for these credits.[36] In 2008, the Energy Improvement and Extension Act granted another $1.5 billion for clean coal tax credits.[37]

Besides tax credits, the federal government has also reclassified royalty payments as capital gains income. Under this reclassification, the coal industry can benefit because capital gains are taxed at a lower rate than other income. Coal companies pay less in taxes because this reclassification gives them a privilege. The federal government has a long-standing practice of tax reclassifications to give special treatment to politically advantaged industries. Beginning in the 1950s, the federal government adopted its first tax reclassifications to encourage coal production. In just the years from 2000 to 2009, $1.3 billion in total deductions have resulted from this reclassification.[38]

Much of America's coal is located on federally owned land, and over the years, different management agencies have made coal extraction on federal lands easier in some cases, but more difficult in others. When a government agency, such as the Bureau of Land Management (BLM), allows a company to mine for coal on federal land, the agency sells a lease to the company. The lease is essentially a contract between the government and the company for a certain period of time and with certain stipulations for each side. If the company violates the terms of the lease, the government can shut down the operation and terminate the lease. If the agency violates the terms of the lease, the company can sue the agency. Leases to federal lands are much more common in the western United States, where the vast majority of federal land is located.

Coal leases from the BLM have become increasingly controversial in recent years because the leases seem to favor politically connected coal companies. When a company wants to mine for coal on BLM land, they must submit an application to the BLM. If BLM officials find the application satisfactory, the agency allows all companies who had approved applications to participate in a lease sale. The BLM arranges the lease sale as a competitive sealed bid procedure, where the highest bidder gets the lease. The highest bidder can only get the lease if the bid is above the fair market value, which BLM officials set.

Many people, including those in the Department of the Interior's Office of the Inspector General (OIG) wanted to know if coal leases on public lands were actually competitive and making money.[39] In 2013, the Government Accountability Office (GAO) found that the BLM's process for determining the "fair market value" may not be accurate. The GAO concluded that a lack of competition, unsatisfactory consideration of export value, and lack of adherence to regulatory standards were skewing the fair market valuation of BLM leases.[40]

The current system of coal leases on BLM lands have become, in effect, a subsidy that allows coal companies to pay less than they would in an actual competitive market.[41] The lack of competition in the bidding process for leases is one of the biggest problems facing the BLM. For example, coal lease sales in the Powder River Basin of Wyoming and Montana have been highly uncompetitive. The Powder River Basin is one of the largest coal mining regions in the United States, supplying 40 percent of the country's coal. In the past 20 years, however, approximately 80 percent of lease sales in the Powder River Basin only received one bid. To make matters worse, no coal lease in the Powder River Basin in this time frame received over two bids. Lease sales are not competitive because so few people are bidding on them, and the determination of a fair market value is unreliable at best, completely inaccurate at worst. The lease system undervalues coal, which means that coal producers can provide coal at a lower cost than we would expect.

Another way that the federal government helps the coal industry is through grants. Most grants, however, are only given to the coal industry for research and development in cleaner and more efficient ways to use coal for electricity. These grants are meant to help coal companies discover better technology more quickly. From 1950 to 2009, the federal government gave almost $34 billion in grants to the coal industry for research and development.[42]

Despite their good intentions, the federal government's grant programs have not always provided successful outcomes. In the late 1980s, the Department of Energy created the Clean Coal Power Initiative (CCPI) to allocate funding for environmentally friendly technologies. At first, the CCPI was mainly focused on addressing acid rain, but by 2000, the CCPI began to focus more heavily on climate change and human health.[43] Since 2003, the CCPI has given out $1.8 billion to coal companies.[44] Between 2003 and 2010, 18 clean coal projects received CCPI funding, but only 4 were successfully completed. Despite hundreds of millions of dollars, the other 14 CCPI recipients have either been discontinued or are still under construction.[45] These failed projects have been a waste of taxpayer dollars.

States Battle over Coal

Besides federal policies, states have also enacted policies that affect how much coal is produced and used to generate electricity. Each state's coal-related policies vary widely. In many states, coal is the largest source of electricity, and the industry employs a significant part of the workforce. In other states, coal is actively discouraged in favor of renewable energy sources. Places that are more dependent on coal for electricity and economic well-being generally have more lenient policies, while other places that rely less on coal have more stringent policies.

Coal is one of the most cost-effective energy sources, and states without much coal-generated electricity pay higher rates.[46] For example, some of the most coal-dependent states are West Virginia, Kentucky, and Wyoming, and these same states have some of the lowest electricity rates in the country.[47] New York and Connecticut, on the other hand, have relatively little coal-generated electricity and also have the highest electricity prices in the nation.[48]

Many state government officials fear the encroachment of federal policies onto their electricity markets. Federal policies generally increase costs due to investments for compliance, and state policy makers fear that federal policies will make electricity rates higher in their states and also hurt industries that employ large portions of their constituency. Low electricity rates are important for local economic health because electricity is essential for almost every industry, especially manufacturing.[49]

Coal-dependent states like Wyoming want to make sure that federal regulations will not make electricity more expensive or subvert industries that help local economies. Wyoming receives about 11.2 percent of its state revenue from the coal industry.[50] Coal is the second largest source of revenue for Wyoming's state and local governments.[51] In 2015, Wyoming joined 14 other states to challenge the EPA's Clean Power Plan, which would impose strict standards on the fossil fuel industry across the country. If the Clean Power Plan were put into effect, Wyoming would have to reduce its carbon dioxide output rate to 56 percent of 2012 levels by 2030.[52] Some estimates suggest that the Clean Power Plan could cause Wyoming's coal production to fall 20 to 45 percent by 2030, which could devastate local employment and economic conditions.[53]

Once again, the political battle between economic prosperity and environmental health is playing out between the states. Other states support the EPA's Clean Power Plan as a way to alleviate the problems of climate change. The governments of 14 states and two cities have stated that they are willing to defend the Clean Power Plan in court.[54] The states in favor of the Clean Power Plan use little to no coal for electricity generation. Eleven of the 14 states that said they would defend the Clean Power Plan produce relatively little coal-fired electricity.

California, for example, supports the Clean Power Plan and has also enacted strict state policies to drive the coal industry out of the state.[55] In 2006, California's state legislature passed the Emissions Performance Standard to forbid long-term investments in coal power plants that do not meet the designated requirements for carbon emissions. Investments in coal power plants then plummeted.[56]

Coal has both economic benefits and environmental impacts. To limit the environmental impacts, government at the state and federal level has tried to reduce coal-generated electricity. But these governments have also enacted policies to make coal even more economical. Lawmakers and regulators create both environmental and economic policies without knowing exactly how their policies will affect average citizens and the environment. Thinking economically, it makes sense why states that generate a large proportion of their electricity from coal and have lower electricity rates generally oppose stricter regulations on coal. States that generate relatively less electricity from coal support stricter regulations on average.

As one of the least expensive and most reliable electricity sources, coal has obvious advantages, but the environmental impacts have made it unpopular in the eyes of many Americans and policy makers. Renewable energy alternatives, however, are often more expensive and less reliable. Each person has different values and different preferences. Some argue that strict regulations are a necessity. Others argue that subsidies produce the best outcomes for the coal industry, taxpayers, or consumers of electricity. These opposing ideas are subjective, and the battle over subjective preferences happens in the political arena. Politics is really a fight over who can get their personal preferences enshrined into law. Contradictory policies regarding coal at both the federal and state levels show that both sides of the political battle surrounding coal have been successful at lobbying for policies that benefit them.

Contradictory government policies drive up costs for both ratepayers and taxpayers. These costs are often unseen because regulations make it more expensive for electricity producers, and these higher costs are passed along to ratepayers. Subsidies that favor the coal industry are costs that are passed along to taxpayers. As long as governments intervene in the energy markets, groups will fight to be chosen as the winner and pass the costs along to someone else.

CHAPTER ELEVEN

Natural Gas

New York's Fracking Fight

In June of 2015, the New York Department of Environmental Conservation and Governor Andrew Cuomo approved a ban on all hydraulic fracking in the state, even though part of New York sits atop the natural gas–rich area called the Marcellus Formation. A few other states have either banned or placed a moratorium on fracking, but New York is the only state to ban fracking that has large fracking potential. The ban came after a seven-year study by the New York Department of Environmental Conservation (DEC). Upon completing its study, the department issued a final report stating, "In the end, there are no feasible or prudent alternatives that would adequately avoid or minimize adverse environmental impacts and that address the scientific uncertainties and risks to public health from this activity. The Department's chosen alternative to prohibit high-volume hydraulic fracturing is the best alternative based on the balance between protection of the environment and public health and economic and social considerations."[1] The DEC concluded that hydraulic fracturing should be banned until its impacts are more fully understood.[2]

The majority of New York's electricity comes from natural gas, meaning that the ban severely restricts local natural gas supplies, as well as economic activity and jobs that would otherwise stay in New York. Instead, New York imports natural gas from other places, including nearby Pennsylvania, where the natural gas industry is booming because of fracking.[3] DEC commissioner Joe Martens advocated for the ban:

> After years of exhaustive research and examination of the science and facts, prohibiting high-volume hydraulic fracturing is the only reasonable

alternative. High-volume hydraulic fracturing poses significant adverse impacts to land, air, water, natural resources and potential significant public health impacts that cannot be adequately mitigated. This decision is consistent with DEC's mission to conserve, improve and protect our state's natural resources, and to enhance the health, safety and welfare of the people of the state.[4]

For the past several years, few environmental topics have been as controversial as hydraulic fracturing, also known as fracking. This technique has allowed the United States to produce vastly larger amounts of oil and natural gas in the past decade than ever before. But, despite cheaper and more abundant energy, as well as economic growth, many people fear that fracking may have significant negative impacts on human health and environmental quality. There is no consensus on the benefits or harms of fracking. Even political leaders in the same party have widely different views regarding fracking, and they have disagreed over the best approach to address concerns over fracking.

The Environmental Protection Agency (EPA), with far more resources than New York's DEC, also studied fracking extensively to determine if there are systematic impacts to human health, especially to drinking water quality. The EPA released a much-anticipated study in 2015, stating, "We did not find evidence that [hydraulic fracturing activities] have led to widespread, systemic impacts on drinking water resources in the United States."[5]

Not only has the EPA found that fracking does not systematically harm drinking water, it also has potential environmental benefits by helping the United States emit less carbon and other types of pollution. Burning coal is dirtier and emits more carbon than natural gas. Thanks to fracking, U.S. supplies of natural gas are larger than ever. The resulting low natural gas prices have provided stiff competition with coal—the nation's largest source of electricity for decades. As coal faces increased competition from natural gas (as well as increasingly stringent regulation that has increased the cost of burning coal for electricity), many power plants have switched from coal to natural gas. Even EPA leaders have admitted that natural gas has helped to reduce emissions overall. In an interview with the Center for American Progress, EPA administrator Gina McCarthy said, "Natural gas, being abundant, has been a game changer in our ability to really move forward with pollution reductions that have been very hard to get our arms around for many decades."[6]

President Obama, although an advocate for renewable energy sources, has expressed his views that domestic natural gas production provides both environmental and economic benefits. Natural gas production in the United States is now tightly linked with fracking and horizontal drilling, and without these technologies, the natural gas industry would be far less productive. President Obama has supported domestic natural gas production as a way to help Americans transition from coal to other renewable sources. In an address at a Gulf Cooperation Council Summit, he said,

> I am working internationally to reduce our carbon emissions and to replace over time fossil fuels with clean energies. Obviously, we start at home with all the work that we've done to, for example, double the use of clean energy. But I think that it is important also to recognize that that is going to be a transition process. In the meantime, we are going to continue to be using fossil fuels. And when it can be done safely and appropriately, U.S. production of oil and natural gas is important. I would rather us—with all the safeguards and standards that we have—be producing our oil and gas, rather than importing it, which is bad for our people, but is also potentially purchased from places that have much lower environmental standards than we do.[7]

In addition to the EPA and the president, the Department of the Interior also opposes fracking bans. Secretary of the Interior Sally Jewell said in early 2015, "There is a lot of misinformation about fracking. I think that localized efforts or statewide efforts in many cases don't understand the science behind it and I think there needs to be more science." She called a fracking ban, "the wrong way to go."[8]

Even other Democratic politicians from New York disagree with Governor Cuomo's support of the DEC ban on fracking. Former New York City mayor Michael Bloomberg criticized the ban, telling the *Wall Street Journal*, "It's a misguided policy. To keep coal-fired power plants in upstate New York and not frack doesn't make any sense at all."[9]

Democratic governors in other states who are also conscious of the same issues as Cuomo have taken dramatically different approaches to fracking. Democratic governor Jerry Brown of California signed a pact to fight global warming with Oregon, Washington, and British Columbia. When asked how he could support the pact and also support fracking, he said,

> The premise of that assertion is that climate change is primarily about fracking. And that's the most absurd idea I've ever heard. This is like saying one tire in the 1 billion vehicles in the world is the problem. Fracking is a point, it's an issue, but climate change is dealing with cars, with power plants, with cows, with agriculture, with cement, with land use, with many, many things. . . . Some of these fracking people, I talked to one young lady, I said,

> "Give science a chance." [She said,] "No, I don't need that, I studied it in college." She did not study fracking the Monterey shale in college, because it has not been studied in any serious way before. But it will, and there will be plenty of time to make comments, the state will have to respond to each of the hundreds of thousands of comments. This is a big deal, and the fact that people have just pretended not to see it as a big deal, shows you the utter emptiness of some of the political debate today. People don't know what the hell they're talking about.[10]

Although federal and state leaders will continue to bicker about the actual effects that fracking has on human health and environmental quality, it is clear that a fracking ban may limit potential economic growth in New York. New York has some of the highest electricity prices in the country.[11] These prices partly reflect the reality that the state is heavily reliant on natural gas for electricity and heating. As of 2016, just over 50 percent of New York's net electricity generation came from natural gas.[12] Since the state produces very little natural gas on its own, it needs to import natural gas from other places. If the fracking ban stays in place, those high prices will not be going away anytime soon.

One of the surprising facts about electricity in New York today is that despite a ban on fracking, the state is actually increasing amounts of natural gas. Fracking in other places, especially Pennsylvania, has made natural gas less expensive, and New York relies on that cheaper natural gas to meet its electricity needs. New York City has also mandated that building owners must burn natural gas instead of oil, which continues to drive up the demand for natural gas.[13] The combination of a fracking ban along with higher demand for natural gas means that New York has been constructing and will continue to construct infrastructure to transport and store natural gas. If the state produced its own gas, this infrastructure would be less necessary.[14]

To make matters worse, the places in New York with the highest fracking potential are also some of the poorest parts of the state. Some areas of New York have threatened to break away from New York State and join Pennsylvania so that they can reap the economic benefits of fracking. Secession is decidedly unlikely, but these threats illustrate the impact some are already feeling from the ban.[15]

New York's ban on fracking makes the state's reliance on natural gas complicated. Unless New Yorkers suddenly decide to stop using electricity, the fracking ban means that electricity suppliers have to either generate electricity from other sources or import natural gas from other states.[16] Importing natural gas often costs more than producing it near the source because natural gas is difficult to store and requires pipeline infrastructure to transport it over long distances. Switching to other energy sources and importing natural gas impose higher costs on electricity producers, and those higher costs are passed along to electricity consumers in the form of higher prices.

The controversial nature of fracking probably will not be resolved anytime soon. Political and economic factors on both sides of the argument make it difficult to know the true costs and benefits of natural gas production. Natural gas production, and fracking in particular, are both highly politicized and have become an emotional issue for many people. As with all energy sources, natural gas and fracking come with trade-offs. Even with more information, the subjective decision of what is "right" and "wrong" will continue to make public policy making a messy process that will come with benefits for one group and costs for another.

Natural Gas: Up-and-Coming Energy Leader

Natural gas is one of the fastest growing electricity sources in the country. In recent years, natural gas developers have combined the technologies of fracking and horizontal drilling, providing access to natural gas reserves far more easily and cheaply than ever before. In addition to these technological advances, government policies have both helped and harmed the natural gas industry. Much like coal, natural gas functions under a contradictory system of federal and state policies that simultaneously boost the industry in some ways while also hindering it with substantial regulation.

Natural gas is a fossil fuel primarily made up of methane, but other flammable hydrocarbons are often mixed in. Natural gas often co-occurs with petroleum, so oil and gas operations are often located in the same place and are extracted at the same time. Once a developer extracts the natural gas from underground, it is processed to remove impurities.

Because natural gas is not dense, it is most easily transported by pipelines. We do not have an effective way to store large quantities of natural gas, and it can be expensive to transport it by truck, train, or ship. To use it most economically, we have to use the gas relatively close to its source, meaning that the markets for gas can vary widely depending on the location. Places close to natural gas wells and refineries, as well as places with large amounts of natural gas infrastructure, can buy and sell gas at much lower prices than those that are farther away. For longer-distance transport, natural gas developers can liquefy or compress natural gas so that it can be shipped by truck, train, or ship. Liquefying or compressing natural gas adds to the costs of transporting natural gas, making long-distance fuel transport less common than other types of energy sources.

Natural gas power plants work to generate electricity in multiple ways. The first way steam generation plants produce electricity is by burning natural gas to boil water into steam, which then turns an electricity-generating turbine. Another type of generation is simple cycle generation, which functions much like a jet engine. Plant operators burn natural gas in a combustion turbine, generating electricity almost immediately. These types of power plants are best equipped to meet the fluctuating levels of peak demand because they are easily ramped up or down. Combined cycle plants incorporate elements from both steam generation plants and simple cycle plants. In combined cycle systems, natural gas is burned to power combustion turbines. Then, the heat from the combustion turbines is transferred to a heat recovery steam generator to power a conventional steam turbine.

The consistency and adaptability to power demand changes are the major advantages of natural gas as an electricity source, which is why many people lobby for favorable policies that make extraction and burning cheaper and easier. At the same time, natural gas has undeniable impacts on the environment and potential effects to human health, which is why many people advocate for fracking bans and strict regulations to restrict how natural gas is extracted and burned. Both sides of the policy fight have been successful at persuading politicians and bureaucrats at the state and federal levels, leading to the contradictory, and sometimes illogical, policies that influence the natural gas industry today.

In the past decade, fracking and horizontal drilling have helped natural gas grow to tie with coal as the top energy source in the United States. In 2015, coal and natural gas each provided about 33 percent of our electricity.[17] Even with fewer subsidies than other energy sources, natural gas has become a leader in the energy sector. The natural gas industry received only 4 percent of all federal subsidies for electricity generation in 2013. In comparison, wind power received about nine times as much in federal subsidies, but it generated less than 5 percent of total U.S. electricity.[18]

Fracking: The Key to Success in the Natural Gas Industry

The key to the growth of the natural gas industry has been the technological innovation of combining fracking and horizontal drilling. Contrary to popular belief, fracking is not a new technology. The earliest forms of fracking date from 1865.[19] In the 1980s, oil and gas developers began pairing fracking and horizontal drilling, which made natural gas competitive with coal.[20] Modern fracking and horizontal drilling is a fairly straightforward process. Developers drill a well down vertically until they reach the geological layer that contains oil and natural gas. The vertical well shafts are cased in concrete to make sure that anything that is injected or extracted does not leak into other rock layers or underground water sources. For the most part, wells are drilled into shale rock layers, also called "shale plays," which can be hundreds of miles wide. These shale layers are located thousands of feet underground and are far deeper than the water table.

Once the well reaches the correct geological layer, the drill then turns 90 degrees and begins to drill horizontally. Horizontal drilling allows a single vertical well to be used for many horizontal shafts 360 degrees around the vertical well shaft.[21] Developers then place explosives deep underground in these horizontal tunnels to crack open rocks to extract the trapped fossil fuels. After using the explosives, natural gas developers inject a mixture of water, sand, and chemicals to prop open the cracks so that oil and gas can escape. The oil and natural gas then flow through the horizontal tunnels and then up the vertical well shaft. Developers then separate the mixture of oil, natural gas, and injection water. The different materials are then sent to refineries so that consumers can use them. The leftover mixture of water, sand, and chemicals is permanently reinjected deep underground, placed in holding ponds at the surface, or recycled to be used again for fracking.[22]

Recent innovations in fracking methods, seismic imaging technology, among other improvements, have allowed energy producers to extract more oil and natural gas more cheaply and easily than they ever have before.[23] Using 2013 data, the Brookings Institute analyzed how much fracking had lowered the cost of natural gas. The institute found that the price of natural gas in the United States was 47 percent below what it would have otherwise been without fracking. According to a Brookings's study, Americans spent $13 billion per year less on their residential gas bills from 2007 to 2013 because of the increased supply of natural gas from fracking.[24] Reuters also analyzed how cheaper natural gas has helped the U.S. economy and found that the natural gas boom increased gross domestic product from manufacturing by 15 percent since 2008.[25] These numbers, however, do not account for environmental impacts and represent only one side of the benefit cost calculation.

The Federal Government's Push for Clean Fossil Fuel

Federal natural gas policies, including both regulations and subsidies, distort the energy market. For example, the federal government enacted price ceilings for natural gas in 1954, which kept natural gas prices artificially cheap. As discussed in Chapter 3, prices in a free market communicate the relative scarcity of a good. When something becomes more expensive, it usually means that it has become scarcer. People had no way to know how scarce natural gas was under the price ceiling, and they demanded more than producers were able to supply at the artificially low price. By the 1970s, nationwide natural gas shortages impacted industry and average citizens across the country negatively. Trying to clean up its own mess, Congress passed the Natural Gas Policy Act in 1978 to deregulate natural gas prices.[26] Without a price ceiling, the price of natural gas more accurately reflects its abundance or scarcity.

Despite this deregulation, the federal government has been financially supporting the natural gas industry for decades, through tax expenditures or direct expenditures. Tax expenditures are a type of tax break, and companies who receive tax expenditures get to pay less in taxes than they otherwise would have. Direct expenditures are cash payments. These payments can take the form of loans, loan guarantees, subsidies, and grants for research and development.[27]

Although renewable energy sources received the largest amount of subsidies in recent years, the fossil fuel industry has historically been the biggest recipient of federal financial support. In 2010 dollars, the federal government has given the equivalent of $837 billion to all energy sources from 1950 to 2010. Of that $837 billion, fossil fuels have received about 70 percent, or $594 billion.[28] Over these 60 years, natural gas received 14 percent of all federal energy incentives, equaling about $121 billion in 2010 dollars. Most of this financial support came in the form of preferential tax policies.[29] Unlike subsidies for renewable energy sources, most subsidies for fossil fuels are permanent and do not have to be renewed.[30]

The natural gas industry has relied on preferential tax policies for several decades. For natural gas, 87 percent of federal expenditures have come in the form of favorable tax policies. In comparison, 60 percent of federal expenditures to renewable energy sources were tax policies, and 33 percent to coal.[31] Under these policies, natural gas producers can lower the amount of taxes they owe if they comply with different government standards. These deductions allow companies to keep a larger portion of their profits, and subsidized production.

Government-granted privileges to the oil and gas industry have been decreasing in recent years. Natural gas and petroleum liquids received only 4 percent of all federal support for electricity in 2013, amounting to roughly $690 million. Of this $690 million, $662 million, or 96 percent, came in the form of tax expenditures.[32] Between 2010 and 2013, tax expenditures to the natural gas and petroleum industry fell roughly $400 million—from $2.7 billion to $2.3 billion.[33] Over the same time period, direct expenditures for natural gas and petroleum products decreased by $18 million.[34]

Although direct and tax expenditures have been decreasing in the past few years, the federal government has been increasing the amount it gives to the oil and gas industry for research and development to encourage the development of cleaner, more efficient, or alternative technologies. Federal support for research and development to the oil and gas industry increased from $9 million to $34 million between 2010 and 2013.[35] The natural gas and petroleum liquids industry also receives loans and loan guarantees from the federal government. The U.S. Department of Agriculture provided $77 million for these loans and loan guarantees in 2010.[36]

Despite these advantages, natural gas faces substantial regulatory constraints that limit its growth. One of the main government agencies that oversees the natural gas industry is the Federal Energy Regulatory Commission (FERC). The natural gas industry must comply with FERC's rules for interstate sales, pipeline construction, regulation enforcement, and environmental concerns.[37] A major hurdle for the natural gas industry is obtaining FERC permits for natural gas pipelines.[38] The Government Accountability Office (GAO) researched how much time it actually takes for natural gas companies to comply with all of FERC's standards for a pipeline permit. The GAO found that it takes 18.6 months, or 558 days, to comply with all requirements for a pipeline permit. If companies choose to skip the prefiling step, they wait an average of 7.5 months, or 225 days, for FERC to review the application.[39]

The problem with unnecessarily long delays in the permitting process is that they can turn potentially profitable projects into wastes of time and money. Not only does the long permitting process hurt natural gas companies, but it can also hurt electricity consumers. New pipelines allow natural gas companies to send natural gas in larger quantities to more places more efficiently. When new pipelines are delayed because of a slow permitting process, the pipes may not be able to keep up with production, which can increase the costs of natural gas. Permits, and the processes put in place to get them, are supposed to make sure that human health and environmental quality are not harmed, but there are trade-offs between safety and economic harm. Limiting the natural gas supply through bureaucratic delays can lead to supply shortages and increased prices.

One federal policy that may be helping the oil and gas industry is the low royalty rates for oil and gas that are extracted on federal land. A royalty rate is the amount that natural gas producers pay back to the government when they extract oil and natural gas on government land. The federal royalty rate has not been updated in almost a century and is much lower than the royalty rates on state-owned land. The 12.5 percent royalty rate on federal lands is half the rate of Texas-owned land. Other natural gas–rich states, including Wyoming, Utah, Montana, and Colorado, all have a royalty rate 16.67 percent on their state lands. Some people argue that the low royalty rates on federal land are a huge revenue loss. The Center for Western Priorities has calculated that taxpayers are losing out on roughly $490 million to $730 million in annual revenue due to low royalty rates on federal lands.[40]

In April 2015, Interior Secretary Sally Jewell said, "It's time to have a candid conversation about whether the American taxpayer is getting the right return for the development of oil and gas resources on public lands. The BLM's regulations have not kept pace with technological advances and market conditions, so this is an important information-gathering step as we seek to improve the way the federal government does business."[41] Her statement shows there has been some momentum to adjust royalty rates.

Also in April 2015, the BLM stated that the agency was seeking comments and suggestions on potential revisions to the royalty rate system.[42] By February 2016, the BLM was still considering some royalty rate hikes. The royalty rate on all existing leases would remain the same, but a proposed rule would raise the royalty rate in some instances. The BLM is also seeking comments on potentially using a fluctuating royalty rate to incentivize producers not to "flare" natural gas.[43] Flaring occurs when oil and gas producers cannot collect or store some natural gas economically, so they burn off excess gas at the well site.[44] Flaring of excess natural gas is wasteful and emits carbon and pollutants without any benefit to society.

Even though it is counterintuitive, flaring natural gas is more environmentally sound than just allowing it to escape into the atmosphere. Flaring undeniably produces harmful emissions, but the emissions from flaring are less harmful than just allowing natural gas to escape into the air. Natural gas is mainly composed of methane, which is a more potent greenhouse gas than carbon dioxide. When methane is released into the atmosphere, it is more efficient at keeping in the sun's radiation than carbon dioxide. When comparing an amount of methane to the same amount of carbon dioxide, methane is 25 times more powerful than carbon dioxide as a greenhouse gas over a 100-year period.[45] Burning natural gas as a flare converts methane into carbon dioxide, which effectively reduces net harm of greenhouse gas emissions.

State Governments and the Fracking Frenzy

In addition to federal laws and regulations, state policies also have hidden implications for taxpayers and electricity consumers. Many states have enacted policies that distort energy markets to make natural gas cheaper or more expensive, including regulations, mandates, and subsidies.

States have the power to regulate fracking because recent federal legislation, especially the Energy Policy Act of 2005, has relegated oversight to the states. In particular, the Energy Policy Act of 2005 shifted federal control over fracking in three key ways. First, the law exempted fracking from the Safe Water Drinking Act of 1974. Second, it exempted oil and gas development from the Clean Water Act's runoff regulations.[46] Third, it changed how the National Environmental Policy Act (NEPA) and the Clean Air Act apply to natural gas drilling.[47] These exemptions prompted state legislatures to form their own natural gas policies.

A majority of states (33 in 2014) in the country produce at least some natural gas. The top natural gas producers are Texas, Pennsylvania, Louisiana, Oklahoma, and Wyoming.[48] Large reservoirs of natural gas are found in geologic formations called "shale plays." One of the largest shale plays is the Marcellus formation, located in the Appalachians of New York, Pennsylvania, Ohio, and West Virginia. Other natural gas–rich areas are found in the Haynesville formation of Louisiana and the Barnett formation in Texas.[49]

Because of the uncertainty and hype that surround fracking, states have taken widely different approaches to fracking regulations. On one extreme end, 3 states have completely banned fracking. An additional 31 states have considered legislation that would ban it. Vermont became the first state to completely ban fracking. In 2012, Governor Peter Shumlin signed the fracking ban into law, despite the fact that the ban was more of a political statement than a purposeful law. When the ban was enacted, Vermont had no known natural gas reserves, and there were no natural gas drilling projects. Shumlin said that the ban was meant to protect Vermont's groundwater if shale gas were ever discovered in the state.[50] Vermont does not produce natural gas and consumes the second smallest amount of natural gas in the nation. It is likely that the ban will not have any economic or environmental impacts because natural gas is a non-issue in Vermont.[51]

Maryland followed Vermont's and New York's examples. In May 2015, Maryland's legislature temporarily banned fracking until October 2017 to alleviate fears that fracking would harm tourism, farming, and water systems. The law that banned fracking also requires the Maryland Department of the Environment to make regulations for future fracking that may take place in the state.[52] If Maryland policy makers allow fracking after the ban expires in 2017, strict regulations may discourage gas companies from drilling in the state.[53] Strict regulations may avoid any potential harms that fracking might inflict on Marylanders, but regulations always come with trade-offs. Potential revenues may be lost and higher electricity prices may result if regulations are too strict.

On the opposite end of the spectrum, Texas policy makers have worked to make sure that the natural gas industry and fracking can happen with as little regulation as possible. Legislators at the state level have limited local governments from banning fracking. In 2015, Texas gave the state complete regulatory authority to oversee the oil and gas industry to avoid regulations at the local level from becoming a patchwork.[54] Patchwork regulations create uncertainty and confusion for producers. Uncertainty and confusion are problematic because they increase the amount of time and effort to comply with varying sets of regulations.[55] Under Texas law, natural gas developers must make arrangements with city officials to drill within city borders, meaning that local governments still have some say within their jurisdictions.[56]

Oklahoma legislators, like those in Texas, limited how local authorities can regulate fracking. Under a 2015 law, local policy makers can only regulate traffic, noise, and odors related to fracking, but they cannot create a blanket ban.[57] Governor Mary Fallin supported this law and justified the state government's actions, saying that local fracking bans would harm the state's economy and create a regulatory patchwork. She explained that if local governments could ban fracking, they "could arbitrarily ban exploration and damage the state's largest industry, largest employers and largest taxpayers."[58]

From complete bans where political favor can be gained at little cost in a state like Vermont, to limited regulation in gas-rich states like Texas and Oklahoma, this patchwork or state regulatory approach illustrates the policy environment that natural gas production faces. As states engage in the regulatory process, the politics of each state influences and molds how fracking policy gets adopted, often with little regard to the evidence.

Environmental Impacts

The wide discrepancy in state-level policies regarding fracking is partly due to the fact that fracking's environmental impacts are difficult to measure and are heavily debated. Two of the biggest concerns with fracking are the disposal of wastewater and the potential contamination of groundwater. The storage and disposal of wastewater is a pressing concern because of the amount of water that is used in the fracking process. Each time a well is fracked, between one and five million gallons of water are pumped into the ground and then pumped back to the surface again.[59] Once fracking water comes back to the surface, oil and gas producers have three options. They can store the contaminated water, treat it, or reinject it back underground. Storing waste water at the surface can potentially cause environmental contamination if the water containment approaches spill. Treating wastewater can also be problematic because many municipal and private water treatment facilities are not capable of treating fracking wastewater.[60]

Most oil and gas developers reinject wastewater into disposal wells located thousands of feet underground. Many people fear that injection wells lubricate faults and spark earthquakes.[61] Areas in Oklahoma have reported increased seismic activity that some scientists have ascribed to improperly injecting wastewater.[62] Wastewater recycling technology is improving, however, which may provide natural gas developers with an alternative to disposal wells.[63] Most earthquakes caused by injection wells are harmless, but in some instances, these earthquakes have caused damage. For example, in November 2016, a magnitude 5.0 tremor caused power outages and structural damage to some buildings in Cushing, Oklahoma. The unique geology of Oklahoma and the large number of injection wells appear to be creating more earthquakes in that state than in other places with fracking.[64] At least in Oklahoma, and potentially other states, earthquakes that occasionally cause damage will be a cost imposed on the general public because of injection wells. It is difficult, however, to measure exactly which earthquakes were caused by injection wells or how severe they would have been if injection wells did not exist.

Groundwater contamination may be the most controversial aspect of fracking, as well as the aspect with the most misinformation. Some researchers have found that many drinking-water wells near fracking operations contain methane. A Duke University study found that methane was in "115 of 141 shallow, residential drinking-water wells" that were within a mile of a fracking well than those more than a mile away.[65]

The EPA conducted an investigation on fracking's impacts to drinking water resources. In 2015, the EPA released the findings of its draft study, which stated: "From our assessment, we conclude there are above and below ground mechanisms by which hydraulic fracturing activities have the potential to impact drinking water resources. These mechanisms include water withdrawals in times of, or in areas with, low water availability; spills of hydraulic fracturing fluids and produced water; fracturing directly into underground drinking water resources; below ground migration of liquids and gases; and inadequate treatment and discharge of wastewater. We did not find evidence that these mechanisms have led to widespread, systemic impacts on drinking water resources in the United States." The EPA conceded that "the number of identified cases where drinking water resources were impacted are small relative to the number of hydraulically fractured wells."[66]

Natural gas development through fracking has trade-offs, like any other energy source. Opponents of fracking argue that it can impact water resources if it is not done properly, and it may induce seismic activity in some places. Proponents respond that the recent fracking boom has been one of the major drivers for reductions in pollution and carbon emissions. They also argue that abundant natural gas has driven economic growth and made electricity cheaper. The pros and cons of natural gas will continue to fuel the political debate that rages over this controversial electricity source.

CHAPTER TWELVE

Conclusion

Electricity is fundamental to daily American life, but the political and economic systems that shape the electricity sector make it complicated and costly. Every electricity source comes with pros and cons, costs and benefits. Even with the best of intentions, government policies that are meant to help electricity consumers or the environment can actually make electricity more expensive, or they can shift the economic burden from electricity bills to April 15th tax bills and do little to improve environmental outcomes.

One of the biggest problems with government intervention is that state and federal officials create electricity policies without knowing the long-range impacts. Politicians and bureaucrats may be intelligent, and their advisers may be even more intelligent, but they are not all-knowing or all-powerful. Good intentions do not always amount to good outcomes, and electricity policy is not exempt from this unfortunate reality. If we measured government policies only by the outcomes they produce, rather than the intentions on which they were created, then we would probably have far fewer policies than we do now. We may also have access to lower electricity prices and a more reliable electrical grid.

Unfortunately, not all politicians and bureaucrats have noble intentions when they form laws and regulations. Sometimes policy makers focus on choosing winners and losers, which in turn helps the policy makers in the form of increased political support from their friends in industry. As rationally self-interested people, government officials choose to do what is best for them. Sometimes what's best for them is what is best for everyone, but most times what is best for policy makers favors a privileged few at the expense of everyone else.

The idea of concentrated benefits and dispersed costs explains why favoritism is consistently enacted and remains over the longer term. When special interest groups have an opportunity to use the government's power, they lobby politicians and influence bureaucrats to make sure that they get what they want. Special interest groups can offer money and support to politicians in exchange for policies that benefit them, or they can make reelection more difficult for those in government when they refuse to help. Laws and regulations that favor certain companies or industries give them concentrated benefits, so it is in their interest to continue to lobby so that those privileges are not taken away. Everyone else, however, is forced to pay for the privileged few to keep the benefits. Most people do not fight back against this privilege because the costs are so dispersed, and the time it would take to fight the policy is too large. When hundreds of millions of people pay taxes, the amount that one person is forced to individually pay for one industry's benefits is relatively small. For the most part, it would cost more for a person to engage in the political process to get privileges revoked than just to stay quiet and pay taxes.

Governments at the state and federal level subsidize and mandate certain forms of energy while also restricting other forms through regulations. For fossil fuel industries, federal and state policies simultaneously encourage and discourage the use of coal and natural gas. These contradictory policies impose higher costs on taxpayers and consumers with few benefits. Contradicting policies also provide evidence that special interest groups with widely different interests have been successful at capturing the legislative and rulemaking processes.

Providing cover for policy makers and pressure to enact these policies are the unique Baptists-and-bootleggers relationships that have emerged in the energy sector. On the one hand, renewable energy companies and environmental groups have come together to demand additional support for their preferred policies of subsidies and mandates for the use of renewable power sources. Renewable energy companies have skillfully used the moral cover of environmental groups to facilitate their own interests. On the other hand, more traditional energy producers have aligned with those interested in protecting economic development and have effectively used that cover to facilitate their preferred outcomes. These relationships coupled with the public choice realities politicians and bureaucrats face have led to an energy policy that simultaneously encourages and discourages nearly every part of the electricity production sector.

For nuclear power, lingering Cold War sentiments and fears of a nuclear meltdown have caused policy makers to leave outdated laws on the books and to severely restrict the industry. America's nuclear infrastructure is aging, and few nuclear power plants are being built to replace the older ones that are being taken off-line. The stagnation in the nuclear industry is largely based on the stringent regulations that make building nuclear power plants unduly costly in both time and money. On top of a stagnating industry, the United States has a nuclear waste problem, but it is largely a political problem, not a technological one. Laws restrict how we can reuse or recycle nuclear waste, and political bickering has halted any progress on a permanent repository for nuclear waste.

Today's political favorites are largely renewable energy sources. In particular, wind and solar have become political darlings. Unfortunately, federal and state policies encourage, subsidize, or mandate most renewable energy sources without considering all the shortcomings or implications. Wind and solar power are not reliable sources of electricity because they are variable and intermittent, and both industries have become dependent on government policies to keep them afloat. Less well-known renewables also benefit from favorable policies, but the outcomes from these policies are not always positive. At industrial scales, biomass can have huge environmental impacts and may actually exacerbate carbon emissions. The geothermal industry is also heavily dependent on government subsidies and loan guarantees, which imposes costs on taxpayers.

Hydropower is the biggest source of renewable energy in the country, but federal and state policies have restricted how much we can take advantage of America's abundant hydro resources. Although mega-dams like Hoover Dam are the most common image of hydropower, the hydropower industry could expand by retrofitting the tens of thousands of smaller non-powered dams across the country. Hydropower developers could also retrofit conduit-based systems, like municipal water pipes and canals, with hydroelectric turbines that would generate electricity with minimal environmental impacts. Recent laws have tried to alleviate the regulatory burden that hinders growth in the industry, but these laws have not gone far enough to tackle the dozens of laws and agencies that oversee hydropower production.

Despite all the issues that electricity policies have caused, people are becoming more concerned with the environment and climate change. As fears about environmental quality and climate change grow stronger, politicians and bureaucrats are likely to enact more laws and pass more regulations. Renewable energy companies will continue to lobby for more subsidies and more mandates. Environmentalists will continue to lobby for stricter regulations on coal and natural gas, and fossil fuels companies will lobby to open up more federal land and to relax regulations. All of this lobbying influences government officials to distort energy markets more and more.

More government policies lead to more complexity and more opportunities for government officials to pick winners and losers at the expense of everyone else. Without significant overhauls, the costs associated with electricity policies remain. Luckily, many government officials are working to fix these problems. In many places, policy makers are working to reform electricity policy to make energy markets more competitive and reduce costs that are imposed on consumers and taxpayers.

The political process could distort energy markets even more, or they could make markets more competitive and less costly. Outside of the political arena, entrepreneurs and innovators continue to make newer and better forms of electricity generation, transmission, and consumption. New forms of electricity storage could become marketable on a mass scale in the next several years, which would help overcome the reliability issues with many renewable energy sources. Other technological advances could make fossil fuel extraction and combustion cleaner and more efficient.

People are inventive and ingenious, and bright minds can help us solve our energy problems today, as they have in past centuries. When regulations are not overly restrictive, entrepreneurs and innovators can engage in a process of discovery to help us find electricity that is environmentally friendly, cost-effective, and reliable.

The world of energy is complex, so it is not surprising that government energy policies do not always achieve their intended goals. In an ideal world, the government could produce perfect solutions to the nation's problems. Unfortunately, the government is made of imperfect, biased people, even if most of them have good intentions. The proper role of government is a debate that has lasted for centuries, and it will likely continue for many more. When people consider what the role of government should be in the electricity sector, they should look specifically at how and why policies are made. They should also consider the outcomes of each policy, not just what the policy was supposed to do. Whatever role the government plays, it should be a facilitator of entrepreneurship and innovation, not an obstruction or enabler of particular sources of electricity generation.

Notes

Introduction

1. Simmons, Randy T., Ryan M. Yonk, and Ken Sim. *Nature Unbound: Bureaucracy vs. the Environment* (Oakland, CA: Independent Institute, 2016).

2. The benefits of mutually beneficial trade will be discussed more in Chapter 3.

3. Riley, Bryan. *U.S. Trade Policy Gouges American Sugar Consumers*. Report. The Heritage Foundation, June 5, 2014. http://www.heritage.org/research/reports/2014/06/us-trade-policy-gouges-american-sugar-consumers.

4. Ryan, Alexander. "Good News for Corn, Bad News for You." *US News and World Report,* March 12, 2014. http://www.usnews.com/opinion/economic-intelligence/2014/03/11/obamas-2015-budget-backs-costly-corn-ethanol-subsidies.

Chapter One

1. "Electricity Explained; Electricity in the United States." Home—Energy Explained, Your Guide to Understanding Energy—Energy Information Administration. http://www.eia.gov/energyexplained/index.cfm?page=electricity_in_the_united_states.

2. *2016–2017 Annual Directory and Statistical Report* (Arlington, VA: American Public Power Association, 2016).

3. Thumann, Albert, and Eric A. Woodroof. *Energy Project Financing: Resources and Strategies for Success* (Lilburn, GA: Fairmont Press, 2009), 93.

4. United States of America, Federal Energy Regulatory Commission, Regional Transmission Organizations (Washington, DC: Federal Energy Regulatory Commission, 1999). https://www.ferc.gov/legal/maj-ord-reg/land-docs/RM99-2A.pdf.

5. Blume, Steven Warren. *Electric Power System Basics: For the Nonelectrical Professional* (Hoboken, NJ: Wiley-Interscience, 2007), 198–200.

6. "What Is U.S. Electricity Generation by Energy Source?"—FAQ—U.S. Energy Information Administration (EIA), April 1, 2016. http://www.eia.gov/tools/faqs/faq.cfm?id=427&t=3.

7. "How Much Electricity Does an American Home Use?" FAQ—U.S. Energy Information Administration (EIA), October 18, 2016. https://www.eia.gov/tools/faqs/faq.cfm?id=97&t=3.

8. "Transparent Cost Database." http://en.openei.org/apps/TCDB.

9. Guezuraga, Begoña, Rudolf Zauner, and Werner Pölz. "Life Cycle Assessment of Two Different 2 MW Class Wind Turbines." *Renewable Energy* 37, no. 1 (2012): 37–44. doi:10.1016/j.renene.2011.05.008.

10. Arvesen, A., and E. G. Hertwich. "Assessing the Life Cycle Environmental Impacts of Wind Power: A Review of Present Knowledge and Research Needs." *Renewable and Sustainable Energy Reviews* 16, no. 8 (October 2012): 5994–6006. http://www.sciencedirect.com/science/article/pii/S1364032112004169.

11. Ibid.

12. Hughes, Gordon. *The Performance of Wind Farms in the United Kingdom and Denmark* (London: Renewable Energy Foundation, 2012).

13. "Transparent Cost Database." http://en.openei.org/apps/TCDB.

14. "Levelized Cost and Levelized Avoided Cost of New Generation Resources in the Annual Energy Outlook 2016." *Annual Energy Outlook 2016,* September 15, 2016. Accessed December 12, 2016. http://www.eia.gov/forecasts/aeo/electricity_generation.cfm.

Chapter Two

1. Brown, Robbie. "In Georgia, Some Vote to Stay Dry on Sundays." *The New York Times*, November 11, 2011. http://www.nytimes.com/2011/11/12/us/georgia-or-most-of-it-ends-sunday-ban-on-alcohol-sales.html?_r=0.

2. Rorabaugh, W. J. "The Sons of Temperance in Antebellum Jasper County." *The Georgia Historical Quarterly* 64, no. 3 (1980): 263–79. http://www.jstor.org/stable/40580645.

3. Jordan, Dave. "Greenville County Voters Say Yes to Sunday Liquor Sales." WSPA.com, November 9, 2016. http://wspa.com/2016/11/09/greenville-county-voters-say-yes-to-sunday-liquor-sales.

4. Hampton, Addie. "Voters to Decide on Sunday Alcohol Sales in Multiple Cities." WSPA.com, September 25, 2016. http://wspa.com/2016/10/25/voters-to-decide-on-sunday-alcohol-sales-in-multiple-cities.

5. Southern Baptist Convention. "On Alcohol Use in America." 2006. http://www.sbc.net/resolutions/1156.

6. Yandle, Bruce. "Bootleggers and Baptists: A Conversation with Bruce Yandle." Mercatus Center at George Mason University. http://mercatus.org/video/bootleggers-and-baptists-conversation-bruce-yandle.

7. Rowley, Liz. "Where Is Marijuana Legal in the United States? List of Recreational and Medicinal States." News.MIc, November 10, 2016. https://mic.com

/articles/126303/where-is-marijuana-legal-in-the-united-states-list-of-recreational-and-medicinal-states#.vTyZgMAAa.

8. Yandle, Bruce. "Bootleggers, Baptists, and the Global Warming Battle." *PERC Policy Series PS*, no. 14 (November 1998): 5–7. http://www.perc.org/sites/default/files/ps14.pdf.

9. "Bootleggers and Baptists: A Conversation with Bruce Yandle." Performed by Bruce Yandle, July 12, 2010. http://mercatus.org/video/bootleggers-and-baptists-conversation-bruce-yandle.

10. Ibid.

11. Simmons, Randy T., Ryan Yonk, and Diana Thomas. "Bootleggers, Baptists, and Political Entrepreneurs: Key Players in the Rational Game and Morality Play of Regulatory Politics." *The Independent Review* 15, no. 3 (Winter 2011): 367–81. http://www.independent.org/pdf/tir/tir_15_03_3_simmons.pdf.

12. "Congress and the Public." Gallup.com, September 28, 2016. http://www.gallup.com/poll/1600/congress-public.aspx.

13. Buchanan, James M. "Politics without Romance." In *The Logical Foundations of Constitutional Liberty*, vol. 1, 48–49 (Indianapolis: Liberty Fund, 1999).

14. Comprehensive Regulatory Impact Analysis: The Cornerstone of Regulatory Reform, Testimony before the Senate Committee on Homeland Security and Governmental Affairs Hearing on "Toward a 21st-Century Regulatory System." U.S. Congress (2015) (testimony of Jerry Ellig). http://mercatus.org/sites/default/files/Ellig-SenateHSGACommittee-Testimony.pdf.

15. Muris, Timothy J. "Rules without Reason: The Case of the FTC." *AEI Journal on Government and Society* 6, no. 5 (September/October 1982): 20–26. http://object.cato.org/sites/cato.org/files/serials/files/regulation/1982/9/v6n5-4.pdf.

16. Graham, John D. "Saving Lives through Administrative Law and Economics." *University of Pennsylvania Law Review* 157, no. 2 (December 2008): 395–540. http://www.jstor.org/stable/40380248.

17. McNutt, Patrick. "Chapter 6 Bureaucracy and Government Output." In *The Economics of Public Choice* (Northampton, MA: Edward Elgar Publishing, 2002).

18. "Value of a Statistical Life Changes." Jason Chaffetz to EPA Administrator Gina McCarthy. July 14, 2016. House Government Oversight Committee, Washington, DC. https://oversight.house.gov/wp-content/uploads/2016/07/2016-07-14-JEC-to-McCarthy-EPA-Value-of-Statistical-Life-Changes-and-EPA-Regs-due-7-29.pdf.

19. Gayer, Ted, and W. Kip Viscusi. "Overriding Consumer Preferences with Energy Regulation." Mercatus Center Working Paper, No. 12-21, July 2012. http://mercatus.org/sites/default/files/Energy_regulations_GayerViscusi_WP1221_1.pdf.

20. Comprehensive Regulatory Impact Analysis: The Cornerstone of Regulatory Reform, Testimony before the Senate Committee on Homeland Security and Governmental Affairs Hearing on "Toward a 21st-Century Regulatory System." U.S. Congress (2015) (testimony of Jerry Ellig).

21. On the Human Costs of the US Regulatory System: Should Congress Pressure Agencies to Make Rules Faster? Testimony before the Senate Judiciary

Committee, Subcommittee on Oversight, Federal Rights, and Agency Action, Cong. (2013) (testimony of Patrick A. McLaughlin). http://mercatus.org/sites/default/files/McLaughlin_human-costs_testimony_073013.pdf.

22. Mitchell, Matthew. *The Pathology of Privilege* (Arlington, VA: The Mercatus Center, 2014), 1–3. https://www.mercatus.org/system/files/Mitchell-web.pdf.

23. Ibid.

24. Ibid.

Chapter Three

1. Peltzman, Sam. "The Effects of Automobile Safety Regulation." *Journal of Political Economy* 83, no. 4 (1975): 677–725. doi:10.1086/260352.

2. Munger, Michael. "They Clapped: Can Price-Gouging Laws Prohibit Scarcity?" Library of Economics and Liberty, January 8, 2007. http://www.econlib.org/library/Columns/y2007/Mungergouging.html.

3. Mitchell, Matthew. *The Pathology of Privilege* (Arlington, VA: The Mercatus Center, 2014), 2–3. http://mercatus.org/sites/default/files/Mitchell-web.pdf.

4. Ibid, 3–7.

5. Bastiat, Frederic. *Selected Essays on Political Economy: What Is Seen and What Is Not Seen* (Irvington-on-Hudson, NY: Foundation for Economic Education, 1995).

6. Hayek, Friedrich A. "The Use of Knowledge in Society." *American Economic Review* 35, no. 4 (September 1945): 519–30. http://www.econlib.org/library/Essays/hykKnw1.html.

7. Surowiecki, James. *The Wisdom of Crowds* (New York: Anchor Books, 2005).

8. Mitchell, Matthew. *The Pathology of Privilege* (Arlington, VA: The Mercatus Center, 2014), 1–3.

9. Buchanan, James M. "Politics without Romance." In *The Logical Foundations of Constitutional Liberty,* vol. 1, 48–49 (Indianapolis: Liberty Fund, 1999).

10. Mitchell, Matthew. *The Pathology of Privilege* (Arlington, VA: The Mercatus Center, 2014), 19–20. http://mercatus.org/sites/default/files/Mitchell-web.pdf.

11. Ibid., 21–33.

12. Ibid., 15–17.

13. Stigler, George J. "The Theory of Economic Regulation." *The Bell Journal of Economics and Management Science* 2, no. 1 (Spring 1971): 3–21. doi:10.2307/3003160.

14. Thoma, Mark. "Explainer: What Is 'Moral Hazard'"? CBS News, November 22, 2013. http://www.cbsnews.com/news/explainer-moral-hazard.

15. "The Solyndra Mess." *The New York Times*, November 24, 2011. http://www.nytimes.com/2011/11/25/opinion/the-solyndra-mess.html.

16. Assessing the Department of Energy Loan Guarantee Program, Testimony before the House Committee on Oversight and Government Reform, Cong. (2012) (testimony of Veronique de Rugy). https://www.mercatus.org/publication/assessing-department-energy-loan-guarantee-program.

17. Mitchell, Matthew. *The Pathology of Privilege* (Arlington, VA: The Mercatus Center, 2014), 8–9. http://mercatus.org/sites/default/files/Mitchell-web.pdf.

18. Bryce, Robert. "SunEdison's Subsidy-Fueled Collapse." April 4, 2016. http://www.nationalreview.com/article/433572/sunedison-failure-energy-subsidies-didnt-help.

19. Dawson, John W., and John J. Seater. "Federal Regulation and Aggregate Economic Growth." *Journal of Economic Growth* 18, no. 2 (March 21, 2013): 137–77. doi:10.1007/s10887-013-9088-y.

20. On the Human Costs of the US Regulatory System: Should Congress Pressure Agencies to Make Rules Faster? Testimony before Senate Judiciary Committee, Subcommittee on Oversight, Federal Rights, and Agency Action, Cong. (2013) (testimony of Patrick A. McLaughlin). http://mercatus.org/sites/default/files/McLaughlin_human-costs_testimony_073013.pdf.

21. Mitchell, Matthew D. "The Pathology of Privilege: The Economic Consequences of Government Favoritism." *SSRN Electronic Journal*, July 9, 2012. doi:10.2139/ssrn.2130566.

22. Yang, Chun-Lei. "Rent Seeking, Technology Commitment, and Economic Development." *Journal of Institutional and Theoretical Economics* 154, no. 4 (December 1998): 640–58. http://www.jstor.org/stable/40752099.

23. Lenway, Stefanie, Randall Morck, and Bernard Yeung. "Rent Seeking, Protectionism and Innovation in the American Steel Industry." *The Economic Journal* 106, no. 435 (March 1996): 410. doi:10.2307/2235256.

24. Shane, Scott. "Technological Opportunities and New Firm Creation." *Management Science* 47, no. 2 (February 2001): 205–20. doi:10.1287/mnsc.47.2.205.9837.

25. Mitchell, Matthew. *The Pathology of Privilege* (Arlington, VA: The Mercatus Center, 2014), 8. http://mercatus.org/sites/default/files/Mitchell-web.pdf.

26. Ibid., 15–17.

Chapter Four

1. "Direct Federal Financial Interventions and Subsidies in Energy in Fiscal Year 2013." Analysis & Projections, March 23, 2015. https://www.eia.gov/analysis/requests/subsidy.

2. Shepard, Dwight B. "Cape Wind Project in Jeopardy as Northeast Utilities, National Grid Terminate Contracts." Masslive.com, 2015. http://www.masslive.com/business-news/index.ssf/2015/01/cape_wind_loses_contracts_with_northeast.html.

3. Ibid.

4. Highet, Daniel. "Offshore Wind Not without Controversy." *Newport This Week*, March 10, 2013. http://www.newportthisweek.com/news/2013-10-03/Front_Page/Offshore_Wind_Not_Without_Controversy.html.

5. Kuffner, Alex. "Block Islanders Still Divided on Proposed Deepwater Wind Project." Providencejournal.com, February 24, 2014. http://www.providencejournal.com/article/20140224/news/302249989.

6. Alternate Energy-Related Uses on the Outer Continental Shelf, §§ 43-388 (2005).

7. "Net Generation for Wind, Monthly." Electricity Data Browser. http://www.eia.gov/electricity/data/browser/#/topic/0?agg=1,0,2&fuel=008&geo=vvvvvvvvvvvvo&sec=o3g&linechart=ELEC.GEN.WND-US-99.A~ELEC.GEN.WND-IA-.

8. "Renewable Energy Production Tax Credit: Stability Drives Growth." http://powerofwind.net/powerofwind/ptc2.

9. "Production Tax Credit for Renewable Energy." Union of Concerned Scientists, 2016. http://www.ucsusa.org/clean-energy/increase-renewable-energy/production-tax-credit.

10. "Database of State Incentives for Renewables & Efficiency®—DSIRE." December 22, 2015. http://www.dsireusa.org.

11. Pfotenhauer, Nancy. "Big Wind's Bogus Subsidies." *US News and World Report*, March 12, 2014. http://www.usnews.com/opinion/blogs/nancy-pfotenhauer/2014/05/12/even-warren-buffet-admits-wind-energy-is-a-bad-investment.

12. United States Government Accountability Office, *Wind Energy: Additional Actions Could Help Ensure Effective Use of Federal Financial Support: Report to Congressional Requesters*, vol. 13, 136th ed. (Washington, DC: GAO, 2013), 14–15+. http://gao.gov/assets/660/652957.pdf.

13. Katzenstein, Warren, Emily Fertig, and Jay Apt. "The Variability of Interconnected Wind Plants." *Energy Policy* 38, no. 8 (August 2010): 4400–410. doi:10.1016/j.enpol.2010.03.069.

14. "How Less Became More: Wind, Power and Unintended Consequences in the Colorado Energy Market." Report. Bentek Energy LLC. 33+. http://docs.wind-watch.org/BENTEK-How-Less-Became-More.pdf.

15. Fisher, Travis. "AWEA's Bold Push for More Wind Welfare—IER." Latest Analysis, April 23, 2014. http://instituteforenergyresearch.org/analysis/aweas-bold-push-for-more-wind-welfare.

16. Wald, Matthew L. "Wind Energy Bumps into Power Grid's Limits." *The New York Times*, August 26, 2008, New York ed., A1 sec. http://www.nytimes.com/2008/08/27/business/27grid.html?pagewanted=1&_r=3&adxnnlx=1412892352-wdYLFNnLcP3ozn5auKUX8Q&.

17. Malewitz, Jim. "$7 Billion Wind Power Project Nears Finish." *The Texas Tribune*, October 14, 2013. http://www.highbeam.com/doc/1G1-345591809.html?refid=easy_hf.

18. "Tribpedia: Crez Transmission Lines." *The Texas Tribune*. https://www.texastribune.org/tribpedia/crez-transmission-lines/about.

19. United States of America, Energy Information Administration. "Levelized Cost and Levelized Avoided Cost of New Generation Resources in the Annual Energy Outlook 2016." 7. http://www.eia.gov/outlooks/aeo/pdf/electricity_generation.pdf.

20. "NREL Calculates Emissions and Costs of Power Plant Cycling Necessary for Increased Wind and Solar in the West." News Releases, September 23, 2013. http://www.nrel.gov/news/press/2013/3299.

21. Oates, David Luke, and Paulina Jaramillo. "Production Cost and Air Emissions Impacts of Coal Cycling in Power Systems with Large-scale Wind Penetration." *Environmental Research Letters* 8, no. 2 (2013): 024022. doi:10.1088/1748-9326/8/2/024022.

22. Mills, Andrew, Ryan Wiser, Michael Milligan, and Mark O'Malley. "Comment on "Air Emissions Due to Wind and Solar Power." *Environmental Science & Technology* 43, no. 15 (December 19, 2009): 6106–107. doi:10.1021/es900831b.

23. "NREL Calculates Emissions and Costs of Power Plant Cycling Necessary for Increased Wind and Solar in the West." News Releases, September 23, 2013. http://www.nrel.gov/news/press/2013/3299.

24. Mills, Andrew, Ryan Wiser, Michael Milligan, and Mark O'Malley. "Comment on 'Air Emissions Due to Wind and Solar Power.'" *Environmental Science & Technology* 43, no. 15 (December 19, 2009): 6106–107. doi:10.1021/es900831b.

25. "Wind Turbine Interactions with Wildlife and Their Habitat." January 2014. http://energy.gov/sites/prod/files/2015/03/f20/AWWI-Wind-Wildlife-Interactions-Factsheet.pdf.

26. Koch, Wendy. "Wind Turbines Kill Fewer Birds Than Do Cats, Cell Towers." *USA Today,* September 15, 2015. http://www.usatoday.com/story/money/business/2014/09/15/wind-turbines-kill-fewer-birds-than-cell-towers-cats/15683843.

27. Hayes, Mark A. "Bats Killed in Large Numbers at United States Wind Energy Facilities." *BioScience* 63, no. 12 (December 2013): 975–79. doi:10.1525/bio.2013.63.12.10.

28. Drouin, Roger. "For the Birds (and the Bats): 8 Ways Wind Power Companies Are Trying to Prevent Deadly Collisions." Grist, January 3, 2014. http://grist.org/climate-energy/for-the-birds-and-the-bats-8-ways-wind-power-companies-are-trying-to-prevent-deadly-collisions.

29. Haapala, Karl R., and Preedanood Prempreeda. "Comparative Life Cycle Assessment of 2.0 MW Wind Turbines." *International Journal of Sustainable Manufacturing* 2, no. 2 (2014): 170–85. doi:http://dx.doi.org/10.1504/IJSM.2014.062496.

30. Dolan, Stacey L., and Garvin A. Heath. "Life Cycle Greenhouse Gas Emissions of Utility-Scale Wind Power." *Journal of Industrial Ecology* 16 (March 30, 2012). doi:10.1111/j.1530-9290.2012.00464.x.

Chapter Five

1. "50 Smartest Companies 2015." *MIT Technology Review,* 2015. https://www.technologyreview.com/lists/companies/2015.

2. Bryce, Robert. "SunEdison's Subsidy-Fueled Collapse." *National Review*, April 4, 2016. Read more at: http://www.nationalreview.com/article/433572/sunedison-failure-energy-subsidies-didnt-help.

3. Mattera, Philip, and Kasia Tarczynska. "Uncle Sam's Favorite Corporations; Identifying the Large Companies That Dominate Federal Subsidies. Rep. Good Jobs First." March 2015. Web.

4. Bryce, Robert. "SunEdison's Subsidy-Fueled Collapse." *National Review*, April 4, 2016. Read more at: http://www.nationalreview.com/article/433572/sunedison-failure-energy-subsidies-didnt-help.

5. "SunEdison Bites the Dust." *The Sustainable Investor*, May 7, 2016. Web. https://thesustainableinvestor.net/2016/05/07/sunedison-bites-the-dust.

6. Ibid.

7. Eckhouse, Brian. "How Wall Street's Favorite Solar Company Spent Itself to the Brink." *Bloomberg Technology*. Bloomberg, March 30, 2016. Web. http://www.bloomberg.com/news/articles/2016-03-30/the-2-6-billion-buying-binge-that-felled-solar-giant-sunedison.

8. Ibid.

9. Feld, Olivia. "How SunEdison Went from Wall Street Star to Bankruptcy." *The Los Angeles Times*. N.p., May 4, 2016. Web.

10. Beckerman, Josh. "SunEdison Gets Justice Department Subpoena, Confirms SEC Inquiry." *The Wall Street Journal: Markets*. Dow Jones, March 31, 2016. Web.

11. Bryce, Robert. "SunEdison's Subsidy-Fueled Collapse." *National Review*, April 4, 2016. Read more at: http://www.nationalreview.com/article/433572/sunedison-failure-energy-subsidies-didnt-help.

12. United States of America. National Renewable Energy Laboratory. "Concentrating Solar Power Basics." NREL, n.d. Web. http://www.nrel.gov/workingwithus/re-csp.html.

13. United States of America. Department of Energy. Office of Energy Efficiency and Renewable Energy. "Linear Concentrator System Basics for Concentrating Solar Power." N.p., August 20, 2013. Web. http://energy.gov/eere/energybasics/articles/linear-concentrator-system-basics-concentrating-solar-power.

14. United States of America. Department of Energy. Office of Energy Efficiency and Renewable Energy. "Dish/Engine System Concentrating Solar Power Basics." N.p., August 20, 2013. Web. http://energy.gov/eere/energybasics/articles/dishengine-system-concentrating-solar-power-basics.

15. United States of America. National Renewable Energy Laboratory. "Concentrating Solar Power Basics." NREL, n.d. Web. http://www.nrel.gov/workingwithus/re-csp.html.

16. "Transparent Cost Database." OpenEI, n.d. Web.

17. CurrentResults.com, Liz Osborn. "Sunniest US Cities—Current Results." N.p., n.d. Web. December 15, 2016. http://www.currentresults.com/Weather-Extremes/US/sunniest-cities.php.

18. "Average Annual Sunshine in American Cities—Current Results." N.p., n.d. Web. December 15, 2016. http://www.currentresults.com/Weather/US/average-annual-sunshine-by-city.php.

19. "Solar Resource Data." PVWatts Calculator. National Renewable Energy Library, n.d. Web. http://pvwatts.nrel.gov/pvwatts.php.

20. Siemens. "Solar Panels Track the Sun for More Efficiency." Phys.org—News and Articles on Science and Technology. N.p., January 11, 2011. Web. http://phys.org/news/2011-01-solar-panels-track-sun-efficiency.html.

21. United States of America. National Renewable Energy Laboratory. "The Impact of Improved Solar Forecasts on Bulk Power System Operations in ISO-NE: 4th Solar Integration Workshop." By Carlo Brancucci Martinez-Anido, Anthony R. Florita, and Bri-Mathias Hodge. 5D00 ed. Vol. CP. Golden, CO: National Renewable Energy Laboratory, 2014. 1. Print. Ser. 62817. http://www.nrel.gov/docs/fy14osti/62817.pdf.

22. United States of America. National Renewable Energy Laboratory. "Investigating the Correlation between Wind and Solar Power Forecast Errors in the Western Interconnection." By Jie Zhang, Bri-Mathis Hodge, and Anthony Florita. Golden: NREL, 2013. 2. Print. http://www.nrel.gov/docs/fy13osti/57816.pdf.

23. International Energy Agency. Photovoltaic Power Systems Programme. "Photovoltaic and Solar Forecasting: State of the Art." By Sophie Pelland, Jan Remnund, Jan Kleissl, Takashi Oozes, and Karel De Brander. Paris: IEA, 2013. Print.

24. U.S. Department of Energy. "The Role of Large Balancing Areas in Integrating Solar Generation." 2011. Web. http://www1.eere.energy.gov/solar/pdfs/50059.pdf.

25. National Renewable Energy Laboratory. "NREL Calculates Emissions and Costs of Power Plant Cycling Necessary for Increased Wind and Solar in the West." September 24, 2013. Web. http://energy.gov/eere/articles/energy-department-report-calculates-emissions-and-costs-power-plant-cycling-necessary.

26. Ibid.

27. Trabish, H. K. "The 'Epic Fail' on Solar's Doorstep—and How the Grid Can Help." October 22, 2014. Web. Accessed December 28, 2016. http://www.utilitydive.com/news/the-epic-fail-on-solars-doorstepand-how-the-grid-can-help/324411.

28. Howarth, D., and B. Monsen. "Renewable Energy Faces Daytime Curtailment in California." Accessed December 28, 2016. http://www.nawindpower.com/issues/NAW1412/FEAT_04_Renewable-Energy-Faces-Daytime-Curtailment-In-California.html.

29. Trabish, H. K. "The 'Epic Fail' on Solar's Doorstep—and How the Grid Can Help." October 22, 2014. Web. Accessed December 28, 2016. http://www.utilitydive.com/news/the-epic-fail-on-solars-doorstepand-how-the-grid-can-help/324411.

30. Ibid.

31. Smith, O., and M. Bell. "Renewables' Bird Problem." October 29, 2013. Accessed on http://blog.rmi.org/blog_2013_10_29_renewables_bird_problem.

32. U.S. Energy Information Administration. "Electricity Storage Technologies Can Be Used for Energy Management and Power Quality." December 14, 2011. http://www.eia.gov/todayinenergy/detail.cfm?id=4310.

33. Ibid.

34. American Physical Society. "Integrating Renewable Electricity on the Grid." 19. Accessed December 28, 2016. Web. http://www.aps.org/policy/reports/popa-reports/upload/integratingelec.pdf.

35. St. John, J. "California Passes Huge Grid Energy Storage Mandate." October 27, 2013. http://www.greentechmedia.com/articles/read/california-passes-huge-grid-energy-storage-mandate.

36. Macgill, Bobby. "California Takes Lead in Developing Energy Storage." *Discovery News*. Accessed January 15, 2015. http://news.discovery.com/earth/california-takes-lead-in-developing-energy-storage-150115.htm.

37. Halper, E. "Power Struggle: Green Energy versus a Grid That's Not Ready." *Los Angeles Times*. Accessed December 2, 2013. http://articles.latimes.com/2013/dec/02/nation/la-na-grid-renewables-20131203#axzz2mXIfGQrx/

38. U.S. Energy Information Administration (EIA). "Direct Federal Financial Interventions and Subsidies in Energy in Fiscal Year 2013." xix and xxi. Accessed December 28, 2016. http://www.eia.gov/analysis/requests/subsidy/pdf/subsidy.pdf.

39. U.S. Energy Information Administration. "Levelized Cost and Levelized Avoided Cost of New Generation Resources in the Annual Energy Outlook 2015." June 3, 2015. Accessed December 28, 2016. http://www.eia.gov/forecasts/aeo/electricity_generation.cfm.

40. U.S. Department of Energy. "Renewable Portfolio Standards (RPS) with Solar or Distributed Generation Provisions." *DSIRE,* March 2015. Accessed December 26, 2016. http://ncsolarcen-prod.s3.amazonaws.com/wp-content/uploads/2015/01/RPS-carveout-map1.pdf.

41. U.S. Department of Energy. "Renewable Portfolio Standard." *DSIRE,* May 20, 2015. Accessed December 28, 2016. http://programs.dsireusa.org/system/program/detail/564.

42. National Renewable Energy Laboratory. "Renewable Portfolio Standards." Accessed December 16, 2016. http://www.nrel.gov/tech_deployment/state_local_governments/basics_portfolio_standards.html.

43. Institute for Energy Research. "The Status of Renewable Electricity Mandates in the States." 25. Accessed December 28, 2016. http://instituteforenergyresearch.org/wp-content/uploads/2011/01/IER-RPS-Study-Final.pdf.

44. Ibid., 50.

45. Sarzynski, A. "The Impact of Solar Incentive Programs in Ten States." George Washington Institute of Public Policy. November 2009. Accessed December 28, 2016. http://gwipp.gwu.edu/files/downloads/GWIPP_Impact10.pdf.

46. (To determine this, we used the EPA's greenhouse gas equivalencies calculator. We entered the 6,100,000 metric tons of CO_2 into the calculator's emissions data. The calculator determined that this would be the equivalent of the emissions generated by 1.6 coal-fired power plants in one year.) United States

Environmental Protection Agency. "Greenhouse Gas Equivalencies Calculator." Accessed December 26, 2016. http://www.epa.gov/cleanenergy/energy-resources/calculator.html#results.

47. Database of State Incentives for Renewables and Efficiency (DSIRE). June 2015. "Tax Credit for Solar Energy Systems on Residential Property (Personal)." U.S. Department of Energy (DOE). Accessed December 28, 2016. http://programs.dsireusa.org/system/program/detail/2636.

48. Adelson, Jeff. "Giving Away Louisiana: Solar Energy Tax credit." December 11, 2014. Accessed December 26, 2016. http://www.theadvocate.com/baton_rouge/news/article_9e1a0dbf-73bf-5c2f-90c6-2fa1fad64a04.html.

49. Adelson, Jeff. "Giving Away Louisiana: Solar Energy Tax Credit." December 11, 2014. Accessed December 26, 2016. http://www.theadvocate.com/baton_rouge/news/article_9e1a0dbf-73bf-5c2f-90c6-2fa1fad64a04.html; DASolar Energy. "Louisiana—Energy Tax Credit, Solar Rebates, and Incentives." 2015. Accessed December 28, 2016. http://www.dasolar.com/energytaxcredit-rebates-grants/louisiana.

50. Adelson, Jeff. "Giving Away Louisiana: Solar Energy Tax Credit." December 11, 2014. Accessed December 26, 2016. http://www.theadvocate.com/baton_rouge/news/article_9e1a0dbf-73bf-5c2f-90c6-2fa1fad64a04.html/.

51. Hughes, J. E., and M. Podolefsky. "Getting Green with Solar Subsidies: Evidence from the California Solar Initiative." May 6, 2014. Accessed December 26, 2016. http://spot.colorado.edu/~jonathug/Jonathan_E._Hughes/Main_files/PV_Subsidies.pdf.

52. Zhang, H., Y. Vorobeychik, J. Letchford, and K. Lakkaraju. "Data-Driven Agent-Based Modeling, with Application to Rooftop Solar Adoption." May 2015. Accessed December 26, 2016. http://vorobeychik.com/2015/abmsolar.pdf.

53. Hughes, J. E., and M. Podolefsky. "Getting Green with Solar Subsidies: Evidence from the California Solar Initiative." May 6, 2014. Accessed December 26, 2016. http://spot.colorado.edu/~jonathug/Jonathan_E._Hughes/Main_files/PV_Subsidies.pdf.

54. Note that a single initiative may contain provisions for more than one energy source and therefore may be counted more than once.

55. United States Government Accountability Office (GAO). "Federal Agencies Implement Hundreds of Initiatives." 17–18. February 1, 2012. Accessed December 26, 2016. http://www.gao.gov/assets/590/588876.pdf.

56. Database of State Incentives for Renewables and Efficiency (DSIRE). "Business Energy Investment Tax Credit (ITC)." October 31, 2014. Accessed December 26, 2016. http://programs.dsireusa.org/system/program/detail/658.

57. George Washington Solar Institute. "What Federal Tax Policies Are Currently in Place for Solar?" Accessed December 26, 2016. http://solar.gwu.edu/q-a/what-federal-tax-policies-are-currently-place-solar.

58. Office of the Comptroller of the Currency. "Public Welfare Investments in Solar Energy Facilities Using Renewable Energy Investment Tax Credit." 1. January 1, 2014. Accessed December 27, 2016. http://www.occ.gov/topics

/community-affairs/publications/fact-sheets/fact-sheet-solar-energy-invest-tax-credits-grants.pdf.

59. United States Government Accountability Office (GAO). "Information on Federal and Other Factors Influencing U.S. Energy Production and Consumption from 2000 through 2013." 76. September 1, 2014. http://www.gao.gov/assets/670/666270.pdf.

60. Solar Energy Industries Association (SEIA). "Solar Investment Tax Credit (ITC)." 2014. http://www.seia.org/policy/finance-tax/solar-investment-tax-credit.

61. Solar Energy Industries Association (SEIA). "Solar Industry Data." Accessed December 26, 2016. http://www.seia.org/research-resources/solar-industry-data.

62. Roth, Sammy. "Solar Slowdown: Stalled Projects Dot Desert Landscape." August 24, 2014. Accessed December 26, 2016. http://www.desertsun.com/story/tech/science/energy/2014/08/24/stalled-solar-projects-desert-landscape/14522553.

63. Natter, Ari. "Solar Industry Launches Lobbying Effort as Tax Deadline Prompts Canceled Projects." *Bloomberg*, October 22, 2014. Accessed December 26, 2016. http://www.bna.com/solar-industry-launches-b17179907013.

64. Consolidated Appropriations Act, 2016. Public Law 114–113. December 18, 2015. Accessed December 16, 2016. https://www.congress.gov/114/plaws/publ113/PLAW-114publ113.pdf.

65. United States Government Accountability Office (GAO). "Information on Federal and Other Factors Influencing U.S. Energy Production and Consumption from 2000 through 2013." 82. (GAO Publication No. GAO-14-836). September 2014. Accessed December 26, 2016. http://www.gao.gov/assets/670/666270.pdf.

66. Mueller, J., and A. Ronen. "Tax Reform, a Looming Threat to a Booming Solar Industry." George Washington Solar Institute. 2. September 2014. Accessed December 26, 2016. http://solar.gwu.edu/file/753/download.

67. George Washington Solar Institute. "What Federal Tax Policies Are Currently in Place for Solar?" Accessed December 26, 2016. http://solar.gwu.edu/q-a/what-federal-tax-policies-are-currently-place-solar.

68. U.S. Partnership for Renewable Energy Finance (US PREF). "MACRS Depreciation and Renewable Energy Finance." 4–5. December 2013. Accessed December 26, 2016. http://www.uspref.org/images/docs/MACRSwhitepaper.pdf.

69. United States Government Accountability Office (GAO). "Information on Federal and Other Factors Influencing U.S. Energy Production and Consumption from 2000 through 2013." 82. GAO Publication No. GAO-14-836. September 2014. Accessed December 26, 2016. http://www.gao.gov/assets/670/666270.pdf.

70. de Rugy, V. "A Guarantee for Failure: Government Lending Under Sed. 1705." Mercatus Center at George Mason University. July 18, 2012. Accessed December 26, 2016. https://science.house.gov/sites/republicans.science.house.gov/files/documents/HHRG-114-SY20-WState-VDeRugy2-20150324.pdf.

71. Marron, Donald. "'Spin Alert' DOE Loans Are Losing Money, Not Making Profits." November 17, 2014. Accessed December 26, 2016. https://dmarron.com/2014/11/17/spin-alert-doe-loans-are-losing-money-not-making-profits.

72. Solar Energy Industries Association (SEIA). "1603 Treasury Program." Accessed December 26, 2016. http://www.seia.org/policy/finance-tax/1603-treasury-program.

73. United States Government Accountability Office (GAO). "Information on Federal and Other Factors Influencing U.S. Energy Production and Consumption from 2000 through 2013." 82. (GAO Publication No. GAO-14-836). September 2014. Accessed December 26, 2016. http://www.gao.gov/assets/670/666270.pdf.

74. U.S. Department of the Treasury. "Overview and Status Update of the §1603 Program." July 31, 2016. Accessed December 26, 2016. http://www.treasury.gov/initiatives/recovery/Documents/STATUS%20OVERVIEW.pdf.

75. U.S. Department of Energy. "Net Metering." *DSIRE,* March 2015. Accessed December 26, 2016. http://ncsolarcen-prod.s3.amazonaws.com/wp-content/uploads/2015/04/Net-Metering-Policies.pdf.

76. If net metering customers were paid wholesale rather than retail electricity prices, they would share the cost of using the electrical grid. This would likely eliminate much of the controversy surrounding net metering programs.

77. Maehlum, Mathias. "What's the Difference between Net Metering and Feed-in Tariffs?" March 15, 2014. http://energyinformative.org/net-metering-feed-in-tariffs-difference.

78. Tanton, Tom. "Reforming Net Metering—Providing a Bright and Equitable future." 7–8. March 2014. American Legislative Exchange Council (ALEC). Accessed December 26, 2016. https://www.alec.org/app/uploads/2015/12/2014-Net-Metering-reform-web.pdf.

79. Energy and Environmental Economics, Inc. "California Net Energy Metering (NEM) Draft Cost-Effectiveness Evaluation." 6. September 26, 2013. Accessed December 26, 2016. http://www.cpuc.ca.gov/NR/rdonlyres/BD9EAD36-7648-430B-A692-8760FA186861/0/CPUCNEMDraftReport92613.pdf.

80. Maehlum, M. 2014, March 15. "What's the Difference between Net Metering and Feed-in Tariffs?" http://energyinformative.org/net-metering-feed-in-tariffs-difference.

81. Clark, Anthony. "As Feed-in Tariff Takes a Backseat, Solar Energy Adjusts." *The Gainesville Sun*, April 26, 2014. Accessed December 26, 2016. http://www.gainesville.com/article/20140426/ARTICLES/140429694?p=1&tc=pg.

82. Electricity Local. "Gainesville, FL Electricity Statistics." Accessed December 28, 2016. http://www.electricitylocal.com/states/florida/gainesville.

83. Clark, Anthony. "As Feed-in Tariff Takes a Backseat, Solar Energy Adjusts." *The Gainesville Sun,* April 26, 2014. Accessed December 26, 2016. http://www.gainesville.com/article/20140426/ARTICLES/140429694?p=1&tc=pg.

84. Ibid.

85. Curry, Christopher. "City Commission Will Not Add to Feed-in Tariff in 2014." *The Gainesville Sun*, December 19, 2013. Accessed December 26, 2016. http://www.gainesville.com/article/20131219/ARTICLES/131219531?p=1&tc=pg.

86. Clark, Anthony. "As Feed-in Tariff Takes a Backseat, Solar Energy Adjusts." *The Gainesville Sun*, April 26, 2014. Accessed December 26, 2016. http://www.gainesville.com/article/20140426/ARTICLES/140429694?p=1&tc=pg.

87. Curry, Christopher. "City Commission Will Not Add to Feed-in Tariff in 2014." *The Gainesville Sun*, December 19, 2013. Accessed December 26, 2016. http://www.gainesville.com/article/20131219/ARTICLES/131219531?p=1&tc=pg.

88. Randazzo, Ryan. "SRP Board OKs Rate Hike, New Fees for Solar Customers." *Azcentral*, February 27, 2015. Accessed December 26, 2016. http://www.azcentral.com/story/money/business/2015/02/26/srp-board-oks-rate-hike-new-fees-solar-customers/24086473.

89. We converted the numbers given in liters to gallons by multiplying the number of liters by 0.264172.

90. Mulvaney, Dustin. "Solar energy Isn't Always as Green as You Think." *IEEE Spectrum*, August 26, 2014. Accessed December 26, 2016. http://spectrum.ieee.org/green-tech/solar/solar-energy-isnt-always-as-green-as-you-think.

91. Keith, Geoff, Sarah Jackson, Alice Napoleon, Tyler Comings, and Jean Ramey. "The Hidden Costs of Electricity." 62. Synapse Energy Economics, Inc., September 19, 2012. http://www.civilsocietyinstitute.org/media/pdfs/091912%20Hidden%20Costs%20of%20Electricity%20report%20FINAL2.pdf.

92. International Energy Agency (IEA). "Water for Energy—Is Energy Becoming a Thirstier Resource?" 10. 2012. http://www.iea.org/media/weowebsite/2012/WEO_2012_Water_Excerpt.pdf.

93. Keith, Geoff, Sarah Jackson, Alice Napoleon, Tyler Comings, and Jean Ramey. "The Hidden Costs of Electricity." 62. Synapse Energy Economics, Inc., September 19, 2012. http://www.civilsocietyinstitute.org/media/pdfs/091912%20Hidden%20Costs%20of%20Electricity%20report%20FINAL2.pdf.

94. Carter, N. T., and Richard J. Campbell. "Water Issues of Concentrating Solar Power (CSP) Electricity in the U.S. Southwest." Congressional Research Service, June 8, 2009. Accessed December 26, 2016. http://www.g-a-l.info/solar-water-use-issues-in-southwest.pdf.

95. Union of Concerned Scientists. "Freshwater Use by US Power Plants." 2. November 2011. http://www.ucsusa.org/sites/default/files/attach/2014/08/ew3-freshwater-use-by-us-power-plants-exec-sum.pdf.

96. Union of Concerned Scientists. "Environmental Impacts of Solar Power." March 5, 2013. http://www.ucsusa.org/clean_energy/our-energy-choices/renewable-energy/environmental-impacts-solar-power.html#.VLBdD2TF-mm; International Renewable Energy Agency. "Concentrating Solar Power." 12. January 2013. http://www.irena.org/DocumentDownloads/Publications/IRENA-ETSAP%20Tech%20Brief%20E10%20Concentrating%20Solar%20Power.pdf.

97. Union of Concerned Scientists. "Environmental Impacts of Solar Power." March 5, 2015. Accessed December 5, 2016. http://www.ucsusa.org/clean_energy/our-energy-choices/renewable-energy/environmental-impacts-solar-power.html.

98. National Renewable Energy Laboratory (NREL). "Impacts on Conventional Generators." January 21, 2014. http://www.nrel.gov/analysis/key_activities_integ_impacts.html.

99. National Renewable Energy Laboratory (NREL). "The Western Wind and Solar Integration Study Phase 2." September 2013. Accessed December 26, 2016. http://www1.eere.energy.gov/wind/pdfs/55588.pdf.

100. Union of Concerned Scientists. "Environmental Impacts of Solar Power." March 5, 2015. Accessed December 5, 2016. http://www.ucsusa.org/clean_energy/our-energy-choices/renewable-energy/environmental-impacts-solar-power.html.

101. Ramzy, Austin. "China's Solar Panel Production Comes at a Dirty Cost." *The New York Times,* June 2, 2014. Accessed December 26, 2016. http://sinosphere.blogs.nytimes.com/2014/06/02/chinas-solar-panel-production-comes-at-a-dirty-cost.

102. Lerner, Louise. "Solar Panel Manufacturing Is Greener in Europe Than China, Study Says." Argonne National Laboratory, May 29, 2014. Accessed December 28, 2016. http://www.anl.gov/articles/solar-panel-manufacturing-greener-europe-china-study-says.

103. Galland, Amy. "Clean & Green: Best Practices in Photovoltaics." 21. 2012. Accessed December 28, 2016. http://www.clca.columbia.edu/Clean&Green-Photovoltaics.pdf.

104. Mulvaney, Dustin. "Solar Energy Isn't Always as Green as You Think." *IEEE Spectrum,* August 26, 2014. Accessed December 26, 2016. http://spectrum.ieee.org/green-tech/solar/solar-energy-isnt-always-as-green-as-you-think.

105. Cha, Ariana Eunjung. "Solar Energy Firms Leave Waste Behind in China." *The Washington Post,* March 9, 2008. Accessed December 28, 2016. http://www.washingtonpost.com/wp-dyn/content/article/2008/03/08/AR2008030802595_pf.html.

106. Dutta, Suparna, and Shakuntala Makhijani. "India's National Solar Mission: A Market Analysis of Phase 1." Worldwatch Institute, November 2012. http://www.worldwatch.org/system/files/WW%20Research%20Note%20India%20Solar%20Market%20Analysis%20Phase%201_FINAL_0.pdf.

107. Moler, Robert. "Protecting Wildlife and Creating Renewable Energy on the Carizzo Plain." Sacramento Fish and Wildlife Office, May 14, 2015. Accessed December 28, 2016. http://www.fws.gov/sacramento/outreach/Featured-Stories/RenewableEnergy-CarrizoPlain/RenewableEnergy-CarrizoPlain.htm.

108. U.S. Fish and Wildlife Service. "Desert Tortoise (Gopherus agassizii)." September 16, 2016. Accessed December 28, 2016. http://ecos.fws.gov/speciesProfile/profile/speciesProfile?spcode=C04L.

109. Wells, Ken. "Where Tortoises and Solar Power Don't Mix." October 10, 2012. Accessed December 26, 2016. http://www.bloomberg.com/bw/articles/2012-10-04/where-tortoises-and-solar-power-dont-mix.

Chapter Six

1. Wainwright, Martin. "In the Shadow of Drax, Not so Much a Fight as a Festival." *The Guardian*, September 1, 2006. https://www.theguardian.com/environment/2006/sep/01/energy.activists.

2. Wainwright, Martin. "Coal Train Ambushed Near Power Station in Climate Change Protest." *The Guardian,* June 14, 2008. https://www.theguardian.com/environment/2008/jun/14/activists.carbonemissions.

3. Ibid.

4. Wainwright, Martin. "Drax Train Hijackers 'Planning to Turn Trial into Second Protest.'" *The Guardian,* June 30, 2009. https://www.theguardian.com/environment/2009/jun/30/drax-train-trial-protest.

5. Ibid.

6. Wainwright, Martin. "Drax Protesters Plead Climate Change Cause to Jury." *The Guardian,* July 1, 2009. https://www.theguardian.com/environment/2009/jul/01/drax-protesters-climate-change-jury.

7. Carter, Helen. "Drax Coal Train Hijackers Sentenced." *The Guardian,* September 4, 2009. https://www.theguardian.com/environment/2009/sep/04/drax-protesters-sentenced.

8. Harrabin, Roger. "Renewable Energy: Burning US trees in UK Power Stations." *BBC*, May 28, 2013. http://www.bbc.com/news/science-environment-22630815.

9. Partnership for Policy Integrity (PFPI). "Carbon Emissions from Burning Biomass for Energy." March 17, 2011. Accessed December 26, 2016. http://www.pfpi.net/carbon-emissions.

10. Schaps, Karolin, and Susanna Twidale. "Drax Shares Fall on UK Plan to Change Biomass Subsidies." *Reuters*, December 12, 2014. Accessed December 26, 2016. http://www.reuters.com/article/2014/12/12/drax-group-stocks-idUSL6N0TW1AW20141212.

11. Morales, Alex. "U.K. Plans Cuts to Solar, Biomass Subsidies to Lower Bills." *Bloomberg Business,* July 22, 2015. http://www.bloomberg.com/news/articles/2015-07-22/u-k-plans-renewable-subsidy-cuts-as-rudd-protects-consumers.

12. Voegele, Erin. "DECC Finalizes Early CfDs, Drax Challenges Subsidy Change." *Biomass Magazine,* April 23, 2014. http://biomassmagazine.com/articles/10307/decc-finalizes-early-cfds-drax-challenges-subsidy-change.

13. Downing, Louise, and Aoife White. "Drax Rises as EU Approves U.K. Biomass Subsidies for MGT Power." *Bloomberg Business*, January 22, 2015. http://www.bloomberg.com/news/articles/2015-01-22/drax-rises-as-eu-approves-u-k-biomass-subsidies-for-mgt-power.

14. Department of Energy & Climate Change. "Controlling the Cost of Renewable Energy." July 22, 2015. https://www.gov.uk/government/news/controlling-the-cost-of-renewable-energy.

15. U.S. Department of Energy. "Xcel Energy Wind and Biomass Generation Mandate." *DSIRE*, February 24, 2015. http://programs.dsireusa.org/system/program/detail/396.

16. U.S. Department of Energy. "Renewable Portfolio Standards." *DSIRE,* 2014. http://www.dsireusa.org/resources/detailed-summary-maps.

17. State of North Carolina Utilities Commission. "Order Modifying the Poultry and Swine Waste Set-Aside Requirements and Granting Other Relief."

(Docket No. E-100, SUB 113). August 28, 2012. http://starw1.ncuc.net/NCUC/ViewFile.aspx?Id=a3c753d4-0820-4961-890a-e594174146a9.

18. U.S. Department of Energy. "Renewable Energy and Energy Efficiency Portfolio Standard." February 3, 2015. http://programs.dsireusa.org/system/program/detail/2660.

19. State of North Carolina Utilities Commission. "Order Modifying the Poultry and Swine Waste Set-Aside Requirements and Granting Other Relief." (Docket No. E-100, SUB 113). August 28, 2012. http://starw1.ncuc.net/NCUC/ViewFile.aspx?Id=a3c753d4-0820-4961-890a-e594174146a9.

20. Taxpayers for Common Sense. "Taxpayer Subsidies for Biomass in the Federal Farm Bill." 4. August 2013. http://www.taxpayer.net/images/uploads/downloads/Taxpayer_Supports_for_Biomass_in_Farm_Bill_Fact_Sheet_FINAL.pdf.

21. Carlton, Jim. "(Bio)Mass Confusion." *The Wall Street Journal*, October 18, 2010. http://www.wsj.com/articles/SB10001424052748704116004575521713188253290.

22. U.S. Department of Energy. "Renewable Electricity Production Tax Credit (PTC)." Accessed December 26, 2016. http://energy.gov/savings/renewable-electricity-production-tax-credit-ptc.

23. Union of Concerned Scientists. "Production Tax Credit for Renewable Energy." 2015. http://www.ucsusa.org/clean_energy/smart-energy-solutions/increase-renewables/production-tax-credit-for.html#.VSTPmxPF-Xx.

24. Kotrba, Ron. "The Power of Association." *Biomass Magazine*. Accessed December 28, 2016. http://biomassmagazine.com/articles/1675/the-power-of-association.

25. Farm Service Agency. "Conservation Fact Sheet." 1. *USDA*, July 2014. https://www.fsa.usda.gov/Internet/FSA_File/bcap_fact_sht_2014.pdf.

26. Farm Service Agency. "USDA Improves Forest Health by Harvesting Biomass for Energy." *USDA*, December 16, 2014. http://www.fsa.usda.gov/FSA/newsReleases?area=newsroom&subject=landing&topic=ner&newstype=newsrel&type=detail&item=nr_20141216_rel_0270.html.

27. Eilperin, Juliet. "The Unintended Ripples from the Biomass Subsidy Program." *The Washington Post*, January 10, 2010. www.washingtonpost.com/wp-dyn/content/article/2010/01/09/AR2010010902023.html.

28. Grooms, Lynn. "Growth of Biomass Industry Dependent on Government Policy." *Farm Industry News*, February 7, 2012. http://farmindustrynews.com/bioenergy/growth-biomass-industry-dependent-government-policy.

29. Bonhag Associates. "North Country Healthcare Systems and Their Biomass CHP System." June 25, 2009. Accessed December 28, 2016. http://web.archive.org/web/20110603004529/http://www.epa.gov/chp/documents/wbnr062509_bonhag.pdf.

30. CONEG Policy Research Center, Inc. & Richmond Energy Associates. "North Country Hospital Biomass Combined Heat and Power Demonstration Project." July 2007. http://www.energycentral.com/download/products/North_Country_Hospital_Biomass.pdf.

31. Lydersen, K. "For Hospitals, Capturing Waste Heat Is a Natural Fit." *Midwest Energy News*, March 26, 2015. http://www.midwestenergynews.com/2015/03/26/for-hospitals-capturing-waste-heat-is-a-natural-fit.

32. Ibid.

33. Clarke, Steve, and Fernando Preto. "Biomass Densification for Energy Production Factsheet." Ontario Ministry of Agriculture, Food and Rural Affairs, June 2001. http://www.omafra.gov.on.ca/english/engineer/facts/11-035.htm.

34. Natural Resources Defense Council (NRDC). "Our Forests Aren't Fuel." March 7, 2016. http://www.nrdc.org/energy/forestsnotfuel.

35. Massachusetts Environmental Energy Alliance. "The Harmful Impacts of Biomass Energy Generation." 2. Accessed December 28, 2016. http://massenvironmentalenergy.org/docs/biomass factsheet from MEEA.pdf.

36. Solid Waste and Infectious Waste Regulations, Ohio Admin. Code § 3745-27-03. Accessed December 28, 2016. http://codes.ohio.gov/oac/3745-27-03.

37. Zeller-Powell, Christine E. "Defining Biomass as a Source of Renewable Energy: The Life-Cycle Carbon Emissions of Biomass Energy, and a Survey and Analysis of Biomass Definitions in States' Renewable Portfolio Standards, Federal Law, and Proposed Legislation." 2011. Accessed December 28, 2016. https://scholarsbank.uoregon.edu/xmlui/bitstream/handle/1794/11483/ZellerPowell__Christine_%20Elizabeth_ms2011sp.pdf?sequence=1.

38. Bergman, R., and J. Zerbe. "Primer on Wood Biomass for Energy." 1. May 26, 2004. http://www.esf.edu/wus/documents/primer_on_wood_biomass_for_energy.pdf.

39. Environmental Protection Agency (EPA). "Non-Hydroelectric Renewable Energy." September 25, 2013. http://www.epa.gov/cleanenergy/energy-and-you/affect/non-hydro.html#biomass.

40. Partnership for Policy Integrity (PFPI). "Carbon Emissions from Burning Biomass for Energy." March 17, 2011. Accessed December 26, 2016. http://www.pfpi.net/carbon-emissions.

41. Walker, Thomas, Peter Cardellichio, Andrea Colnes, John Gunn, Brian Kittler, Bob Perschel, Christopher Recchia, and David Saah. "Biomass Sustainability and Carbon Policy Study." 95. Manomet Center for Conservation Sciences, June 2010. Accessed December 28, 2016. http://www.mass.gov/eea/docs/doer/renewables/biomass/manomet-biomass-report-full-hirez.pdf.

42. Booth, Mary S. "Trees, Trash, and Toxics: How Biomass Energy Has Become the New Coal." 5. April 2, 2014. Accessed December 28, 2016. http://www.pfpi.net/wp-content/uploads/2014/04/PFPI-Biomass-is-the-New-Coal-April-2-2014.pdf.

43. Sheehan, M., S. Chirillo, J. Schlossberg, W. Sammons, M. Leonard, and Energy Justice Network. "Biomass Electricity: Clean Energy Subsidies for a Dirty Industry." 4. June 2011. http://www.pfpi.net/wp-content/uploads/2011/06/BAP-Biomass-Projects-Report.pdf.

44. Massachusetts Environmental Energy Alliance. "The Harmful Impacts of Biomass Energy Generation." 1. Accessed December 28, 2016. http://massenvironmentalenergy.org/docs/biomass factsheet from MEEA.pdf.

45. Partnership for Policy Integrity (PFPI). "Biomass Energy Overview." March 17, 2011. http://www.pfpi.net/biomass-basics-2.

46. U.S. Energy Information Administration (EIA). "Electricity Data Browser." Accessed December 5, 2015. http://www.eia.gov/electricity/data/browser/#/plant/57241?freq=M&start=201309&end=201506&ctype=linechart&ltype=pin&columnchart=ELEC.PLANT.GEN.57241-ALL-ALL.M&linechart=ELEC.PLANT.GEN.57241-ALL-ALL.M&maptype=0&pin=.

47. Florida Department of Environmental Protection. "Gainsville Renewable Energy Center." 10. December 28, 2010. http://www.dep.state.fl.us/air/emission/bioenergy/gainesville/GRECFPermit.pdf.

48. National Association of Conservation Districts. "Woody Biomass Desk Guide & Toolkit: Appendix A." 119. November 1, 2008. http://www.nacdnet.org/wp-content/uploads/2016/06/Woody-Biomass-Desk-Guide-and-Toolkit.compressed.pdf.

49. Used online conversion tool to convert acres to square miles: 1 square mile = 640 acres; 30,500 acres/640 = 47.66 square miles.

50. Wolf, Dale, and David A. Fiske. "Planting and Managing Switchgrass for Forage, Wildlife, and Conservation." May 1, 2009. Accessed December 28, 2016. Virginia Cooperative Extension. http://pubs.ext.vt.edu/418/418-013/418-013.html.

51. Union of Concerned Scientists. "Environmental Impacts of Biomass for Electricity." Accessed 2015. http://web.archive.org/web/20150611221139/http://www.ucsusa.org/clean_energy/our-energy-choices/renewable-energy/environmental-impacts-biomass-for-electricity.html#.WGQP6LYrJTY.

52. Searchinger, Timothy, Ralph Heimlich, R. A. Houghton, Fengxia Dong, Amani Elobeid, Jacinto Fabiosa, Simla Tokgoz, Dermot Hayes, and Tun-Hsiang Yu. "Use of U.S. Croplands for Biofuels Increases Greenhouse Gases through Emissions from Land-Use Change." *Science* 319, no. 5867 (February 29, 2008): 1238–40. doi:10.1126/science.1151861.

53. Agricultural Marketing Resource Center (AgMRC). "Switchgrass Profile." August 2013. Accessed December 28, 2016. http://www.agmrc.org/commodities_products/biomass/switchgrass-profile/.

54. To get these numbers, we multiplied 1,500 acres by 600 megawatts, then converted acres to square miles.

55. Searchinger, Tim, and Ralph Hemlich. "Avoiding Bioenergy Competition for Food Crops and Land." World Resources Institute, January 2015. http://www.wri.org/publication/avoiding-bioenergy-competition-food-crops-and-land.

56. Booth, Mary S. "Trees, Trash, and Toxics: How Biomass Energy Has Become the New Coal." 16. April 2, 2014. Accessed December 28, 2016. http://www.pfpi.net/wp-content/uploads/2014/04/PFPI-Biomass-is-the-New-Coal-April-2-2014.pdf.

57. Biomass Energy Resource Center (BERC). "Grass Energy Basics." Accessed December 28, 2016. http://www.biomasscenter.org/resource-library/fact-sheets/grass-energy-basics.

58. Vermont Grass Energy Partnership. "Technical Assessment of Grass Pellets as Boiler Fuel in Vermont." January 2011. http://www.vsjf.org/assets/files/RFPs/VT%20Grass%20Pellet%20Feasibility%20Study%202010.pdf.

59. Department of Ecology, State of Washington. "Prevention of Significant Deterioration: Getting a Permit." Accessed December 28, 2016. http://www.ecy.wa.gov/programs/air/psd/PSD_PDFS/PSD_getpermit.pdf.

60. Booth, Mary S. "Trees, Trash, and Toxics: How Biomass Energy Has Become the New Coal." 5. April 2, 2014. Accessed December 28, 2016. http://www.pfpi.net/wp-content/uploads/2014/04/PFPI-Biomass-is-the-New-Coal-April-2-2014.pdf.

61. Ibid., 38–40.

62. Ibid., 57–59.

63. Ibid., 8.

Chapter Seven

1. Lipton, Eric, and Clifford Krauss. "A U.S.-Backed Geothermal Plant in Nevada Struggles." *The New York Times,* October 2, 2011. http://www.nytimes.com/2011/10/03/business/a-us-backed-geothermal-plant-in-nevada-struggles.html?_r=0.

2. Neubauer, Chuck. "Lights Go Dim on Another Energy Project; Geothermal Losses Pile Up." *The Washington Times*, July 5, 2012. Accessed 2016. http://www.washingtontimes.com/news/2012/jul/4/lights-go-dim-on-another-energy-project/.

3. Lipton, Eric, and Clifford Krauss. "A U.S.-Backed Geothermal Plant in Nevada Struggles." *The New York Times,* October 2, 2011. http://www.nytimes.com/2011/10/03/business/a-us-backed-geothermal-plant-in-nevada-struggles.html?_r=0.

4. Fehrenbacher, Katie. "A Troubled Geothermal Plant Finds a Savior in a Startup and Vinod Khosla." *Fortune,* May 20, 2015. http://fortune.com/2015/05/20/geothermal-altarock-nevada.

5. Institute for Energy Research (IER). "Encyclopedia Entry: Geothermal." 2014. http://instituteforenergyresearch.org/topics/encyclopedia/geothermal.

6. U.S. Energy Information Administration (EIA). "Geothermal Resources Used to Produce Renewable Electricity in Western States." September 8, 2014. http://www.eia.gov/todayinenergy/detail.cfm?id=17871; Geothermal Energy Association. "2015 Annual U.S. & Global Geothermal Power Production Report [Figure 5: Developing Projects and Capacity by Country or Territory]." 11. February 2015. http://geo-energy.org/reports/2015/2015%20Annual%20US%20%20Global%20Geothermal%20Power%20Production%20Report%20Draft%20final.pdf.

7. Eliasson, E. T., S. Thorhallsson, and B. Steingrimsson. "Geothermal Power Plants. Geothermal Training Programme." 2. United Nations University, March 23, 2014. http://www.os.is/gogn/unu-gtp-sc/UNU-GTP-SC-18-12.pdf.

8. U.S. Department of Energy Office of Energy Efficiency & Renewable Energy. "A History of Geothermal Energy in America." 2013. Accessed December 26, 2016. http://energy.gov/eere/geothermal/history-geothermal-energy-america.

9. California Energy Commission. "Types of Geothermal Power Plants." *Energy Almanac,* 2015. http://energyalmanac.ca.gov/renewables/geothermal/types.html.

10. Eliasson, E. T., S. Thorhallsson, and B. Steingrimsson. "Geothermal Power Plants. Geothermal Training Programme." 2–5. United Nations University, March 23, 2014. http://www.os.is/gogn/unu-gtp-sc/UNU-GTP-SC-18-12.pdf.

11. Ibid.

12. Kubiszewski, I. "Geothermal Steam Act of 1970, United States." September 4, 2008. http://www.eoearth.org/view/article/152963.

13. U.S. Department of Energy Office of Energy Efficiency & Renewable Energy. "A History of Geothermal Energy in America." 2013. Accessed December 26, 2016. http://energy.gov/eere/geothermal/history-geothermal-energy-america.

14. Schochet, Daniel, and John Mock. "How the Department of Energy Loan Guarantee Program Paved the Way for the Growth of the Geothermal Industries." *Geothermal Resources Council Transactions* 18 (October 1994): 61–65.

15. Union of Concerned Scientists. "Public Utility Regulatory Policy Act (PURPA)." Accessed December 26, 2016. http://www.ucsusa.org/clean_energy/smart-energy-solutions/strengthen-policy/public-utility-regulatory.html#.VhXr0npViko.

16. Wallace, Megan. "A Negotiated Alternative to Mandatory Wheeling." *Energy Law Journal* 10, no. 99 (1989): 99–120.

17. Handmaker, Robert. "Deregulating the Transmission of Electricity: Wheeling under P.U.R.P.A. Sections 203, 204, and 205." *Washington University Law Review* 67, no. 2 (January 1989): 435–60.

18. Union of Concerned Scientists. "Public Utility Regulatory Policy Act (PURPA)." Accessed December 26, 2016. http://www.ucsusa.org/clean_energy/smart-energy-solutions/strengthen-policy/public-utility-regulatory.html#.VhXr0npViko.

19. Matek, Benjamin. "2015 Annual U.S. & Global Geothermal Power Production Report [Figure 6: U.S. Industry Geothermal Nameplate & Net Capacity]." Geothermal Energy Association. 12. February 2015. http://geo-energy.org/reports/2015/2015%20Annual%20US%20%20Global%20Geothermal%20Power%20Production%20Report%20Draft%20final.pdf.

20. Elefant, Carolyn. "Reviving PURPA's Purpose: The Limits of Existing State Avoided Cost Ratemaking Methodologies in Supporting Alternative Energy Development and a Proposed Path for Reform." The Law Offices of Carolyn Elefant. 1. http://www.recycled-energy.com/images/uploads/Reviving-PURPA.pdf.

21. Matek, Benjamin. "2015 Annual U.S. & Global Geothermal Power Production Report [Figure 6: U.S. Industry Geothermal Nameplate & Net Capacity]." Geothermal Energy Association. 12. February 2015. http://geo-energy.org/reports/2015/2015%20Annual%20US%20%20Global%20Geothermal%20Power%20Production%20Report%20Draft%20final.pdf.

22. Arhar, B. C. "Geothermal Access to Federal and Tribal Lands: A Progress Report." National Renewable Energy Laboratory. 2. September 2002. http://www.nrel.gov/docs/fy02osti/32204.pdf.

23. Sherlock, Molly F. "The Renewable Electricity Production Tax Credit: In Brief." Congressional Research Service. 3. October 2, 2014. http://cdn.akingump.com/images/content/3/2/v2/32426/CRS-PTC-Report-of-10-2-14-Sherlock.pdf.

24. U.S. Department of Energy. "Renewable Electricity Production Tax Credit (PTC)." Accessed December 26, 2016. http://energy.gov/savings/renewable-electricity-production-tax-credit-ptc.

25. American Council on Renewable Energy (ACORE). "The Outlook for Renewable Energy in America." 21. 2014. http://www.acore.org/files/pdfs/ACORE_Outlook_for_RE_2014.pdf.

26. Matek, Benjamin. "2015 Annual U.S. & Global Geothermal Power Production Report." Geothermal Energy Association. 12. February 2015. http://geo-energy.org/reports/2015/2015%20Annual%20US%20%20Global%20Geothermal%20Power%20Production%20Report%20Draft%20final.pdf.

27. Néron-Bancel, T. "Geothermal Revenue under the Energy Policy Act of 2005: Income Distribution at Federal, State, and County Levels." Geothermal Energy Association. 5. December 14, 2008. http://geo-energy.org/reports/Geothermal_Revenue_Under_the_Energy_Policy_Act_of_2005_Final.pdf.

28. U.S. Department of Energy Office of Energy Efficiency & Renewable Energy. "American Recovery and Reinvestment Act." Accessed December 28, 2016. http://energy.gov/eere/geothermal/american-recovery-and-reinvestment-act.

29. Percentage Change Calculation: (16,628 GWh –14,692 GWh) / 14,692 GWh $\cong$ .13 $\cong$ 13 percent; U.S. Energy Information Administration. (2015, August). "Electric Power Monthly with Data for June 2015 [Table 1.1.A. Net Generation from Renewable Sources: Total (All Sectors), 2005-June 2015]." 16. August 2015. http://www.eia.gov/electricity/monthly/current_year/august2015.pdf.

30. Percentage calculation: $312 million/ $345 million $\cong$ .9 $\cong$ 90%; United States Energy Information Administration (EIA). "Direct Federal Financial Interventions and Subsidies in Energy in Fiscal Year 2013. Analysis & Projections." United States Department of Energy. [Table ES2. Quantified energy-specific subsidies and support by type, FY 2010 and FY 2013 (million 2013 dollars)]. 2015. http://www.eia.gov/analysis/requests/subsidy.

31. DSIRE. "Business Energy Investment Tax Credit (ITC)." United States Department of Energy, December 21, 2015. http://programs.dsireusa.org/system/program/detail/658.

32. United States Energy Information Administration (EIA). "Direct Federal Financial Interventions and Subsidies in Energy in Fiscal Year 2013. Analysis & Projections." United States Department of Energy, 2015. http://www.eia.gov/analysis/requests/subsidy.

33. California Energy Commission. "Geothermal Grant and Loan Program (GRDA) Law and Regulations." Accessed December 28, 2016. http://www.energy.ca.gov/geothermal/grda_regs.html.

34. Ibid.

35. Department of Energy. "Database of State Incentives for Renewables and Efficiency." *DSIRE,* 2016. http://www.dsireusa.org.

36. Ibid.

37. Skinner, Lara. "Solar, Geothermal Guides Nevada to RPS Goals." October 27, 2004. http://www.renewableenergyworld.com/articles/2004/10/solar-geothermal-guides-nevada-to-rps-goals-17444.html.

38. U.S. Department of Energy. "What Is an Enhanced Geothermal System (EGS)?" 1. September 2012. https://www1.eere.energy.gov/geothermal/pdfs/egs_basics.pdf.

39. Matek, Benjamin. "2015 Annual U.S. & Global Geothermal Power Production Report." Geothermal Energy Association. 10–18. February 2015. http://geo-energy.org/reports/2015/2015%20Annual%20US%20%20Global%20Geothermal%20Power%20Production%20Report%20Draft%20final.pdf.

40. Hance, Cédric. "Factors Affecting Costs of Geothermal Power Development." Geothermal Energy Association. 4. August 2005. Accessed December 28, 2016. http://geo-energy.org/reports/Factors%20Affecting%20Cost%20of%20Geothermal%20Power%20Development%20-%20August%202005.pdf.

41. Ibid.

42. International Finance Corporation. "Success of Geothermal Wells: A Global Study." 4–17. June 2013. http://www.ifc.org/wps/wcm/connect/7e5eb4804fe24994b118ff23ff966f85/ifc-drilling-success-report-final.pdf?MOD=AJPERES.

43. Cochener, John. "Quantifying Drilling Efficiency." U.S. Energy Information Administration Office of Integrated Analysis and Forecasting. 7. June 28, 2010. http://www.eia.gov/workingpapers/pdf/drilling_efficiency.pdf.

44. Sandia National Laboratories. "Handbook for Best Practices for Geothermal Drilling." 14. December 2010. http://www1.eere.energy.gov/geothermal/pdfs/drillinghandbook.pdf.

45. Augustine, Chad, Jefferson Tester, Susan Petty, and Bill Livesay. "A Comparison of Geothermal with Oil and Gas Well Drilling Costs." Massachusetts Institute of Technology, Black Mountain Consulting. 7. January 30, 2006. https://pangea.stanford.edu/ERE/pdf/IGAstandard/SGW/2006/augustin.pdf.

46. Schlumberger Business Consulting. "Improving the Economics of Geothermal Development through an Oil and Gas Industry Approach." 5. Accessed December 26, 2016. http://www.slb.com/~/media/Files/business_consulting/industry_articles/thermal_dev.pdf.

47. Ibid., 4.

48. U.S. Energy Information Administration (EIA). "Updated Capital Cost Estimates for Utility Scale Electricity Generating Plants [Table 1]." April 12, 2013. http://www.eia.gov/forecasts/capitalcost.

49. International Energy Agency (IEA). "Renewable Energy Essentials: Geothermal." 3. 2010. https://www.iea.org/publications/freepublications/publication/Geothermal_Essentials.pdf.

50. International Finance Corporation. "Success of Geothermal Wells: A Global Study." 5. June 2013. http://www.ifc.org/wps/wcm/connect/7e5eb4804fe24994b118ff23ff966f85/ifc-drilling-success-report-final.pdf?MOD=AJPERES.

51. U.S. Department of Energy. "Department of Energy Announces Up to $70 Million to Advance Technology and Reduce Cost of Geothermal Energy." June 8, 2011. http://energy.gov/articles/department-energy-announces-70-million-advance-technology-and-reduce-cost-geothermal-energy.

52. Kristmannsdóttir, Hrefna, and Halldór Ármannsson. "Environmental Aspects of Geothermal Energy Utilization." *Geothermics* 32, no. 4–6 (August 2003): 451–61. doi:10.1016/s0375-6505(03)00052-x.

53. United States Department of Energy. "Chapter 8 Environmental Impacts, Attributes, and Feasibility Criteria." Office of Energy Efficiency and Renewable Energy. 8–6. https://www1.eere.energy.gov/geothermal/pdfs/egs_chapter_8.pdf.

54. Kagel, Alyssa, Diana Bates, and Karl Gawell. "A Guide to Geothermal Energy and the Environment." Geothermal Energy Association. 44. April 2007. http://www.geo-energy.org/reports/Environmental%20Guide.pdf.

55. United States Department of Energy. "Chapter 8 Environmental Impacts, Attributes, and Feasibility Criteria." Office of Energy Efficiency and Renewable Energy. 8–6. https://www1.eere.energy.gov/geothermal/pdfs/egs_chapter_8.pdf.

56. Environmental Protection Agency. "Class V UIC Study Fact Sheet: Geothermal Direct Heat Return Flow Wells." Accessed December 28, 2016. https://www.epa.gov/sites/production/files/2015-08/documents/fs_geo_heat_wells.pdf.

57. Macknick, Jordan, Robin Newmark, Garvin Heath, and K. C. Hallett. "A Review of Operational Water Consumption and Withdrawal Factors for Electricity Generating Technologies." National Renewable Energy Laboratory. 12. March 2011. http://www.nrel.gov/docs/fy11osti/50900.pdf.

58. Ibid., 2.

59. Kagel, Alyssa. "The State of Geothermal Technology Part II: Surface Technology." Geothermal Energy Association. 9. January 2008. http://www.geo-energy.org/reports/Geothermal%20Technology%20-%20Part%20II%20%28Surface%29.pdf.

60. Matek, Benjamin. "Promoting Geothermal Energy: Air Emissions Comparison and Externality Analysis." Geothermal Energy Association. 7. April 2013. http://geo-energy.org/events/Air%20Emissions%20Comparison%20and%20Externality%20Analysis_Publication.pdf.

61. Kagel, Alyssa, Diana Bates, and Karl Gawell. "A Guide to Geothermal Energy and the Environment." Geothermal Energy Association. 44. April 2007. http://www.geo-energy.org/reports/Environmental%20Guide.pdf.

62. Matek, Benjamin. "Promoting Geothermal Energy: Air Emissions Comparison and Externality Analysis." Geothermal Energy Association. 7. April 2013. http://geo-energy.org/events/Air%20Emissions%20Comparison%20and%20Externality%20Analysis_Publication.pdf.

63. Sullivan, J. L., C. E. Clark, Yuan Han, and M. Wang. "Life-Cycle Analysis Results of Geothermal Systems in Comparison to Other Power Systems: Part II." Center for Transportation Research. Energy Systems Division, Argonne National Laboratory. 29. November 2011. https://greet.es.anl.gov/files/lca-goethermal.

64. Stephens, Tim. "Geothermal Power Facility Induces Earthquakes, Study Finds." University of California Santa Cruz, July 11, 2013. http://news.ucsc.edu

/2013/07/geothermal-earthquakes.html; Brodsky, E. E., and L. J. Lajoie. "Anthropogenic Seismicity Rates and Operational Parameters at the Salton Sea Geothermal Field." *Science* 341, no. 6145 (July 11, 2013): 543–46. doi:10.1126/science.1239213.

65. Harmon, Katherine. "How Does Geothermal Drilling Trigger Earthquakes?" *Scientific American*, June 29, 2009. http://www.scientificamerican.com/article/geothermal-drilling-earthquakes.

66. Ibid.

Chapter Eight

1. Hansen, Megan, Randy Simmons, and Ryan Yonk. "The Regulatory Noose: Logan City's Adventures in Micro-Hydropower." *Energies* 9, no. 7 (June 23, 2016): 482. doi:10.3390/en9070482.

2. Ibid.

3. Ibid.

4. Abbey, Edward. *The Monkey Wrench Gang* (Philadelphia: Lippincott, 1975).

5. Brekke, Dan. "New Lawsuit Seeks to Force San Francisco to Drain, Restore Hetch Hetchy." *KQED News*, April 27, 2015. http://ww2.kqed.org/news/2015/04/27/new-lawsuit-seeks-to-force-city-to-drain-restore-hetch-hetchy.

6. U.S. Department of Energy Office of Energy Efficiency & Renewable Energy. "Types of Hydropower Plants." Accessed December 29, 2016. http://energy.gov/eere/water/types-hydropower-plants.

7. Center for Climate and Energy Solutions. *Hydropower Climate TechBook.* Accessed December 29, 2016. http://www.c2es.org/docUploads/Hydropower10%2009_FINAL_cleanPDF.pdf.

8. Kosnik, Lea. "The Potential of Water Power in the Fight against Global Warming in the US." *Energy Policy* 36, no. 9 (September 2008): 3252–65. doi:10.1016/j.enpol.2008.05.009.

9. Ibid.

10. Hadjerioua, Boualem, Yaxing Wei, and Shih-Chieh Kao. "An Assessment of Energy Potential at Non-Powered Dams in the United States." Oak Ridge National Laboratory, April 2012. http://nhaap.ornl.gov/sites/default/files/NHAAP_NPD_FY11_Final_Report.pdf.

11. U.S. Department of Energy. "Pumped Storage and Potential Hydropower from Conduits." *Energy.gov*, February 2015. http://energy.gov/sites/prod/files/2015/06/f22/pumped-storage-potential-hydropower-from-conduits-final.pdf.

12. Ibid.

13. Navigant Consulting, Inc. "Statewide Small Hydropower Resource Assessment." California Energy Commission, June 2006. http://www.energy.ca.gov/2006publications/CEC-500-2006-065/CEC-500-2006-065.PDF; U.S. Department of Energy. "Pumped Storage and Potential Hydropower from Conduits." *Energy.gov*, February 2015. http://energy.gov/sites/prod/files/2015/06/f22/pumped-storage-potential-hydropower-from-conduits-final.pdf.

14. U.S. Department of Energy. "Pumped Storage and Potential Hydropower from Conduits." *Energy.gov*, February 2015. http://energy.gov/sites/prod/files/2015/06/f22/pumped-storage-potential-hydropower-from-conduits-final.pdf.

15. Zhang, Qin F., Brennan Smith, and Wei Zhang. "Small Hydropower Cost Reference Model." Oak Ridge National Laboratory, October 2012. http://info.ornl.gov/sites/publications/files/pub39663.pdf.

16. U.S. Department of Energy. "Pumped Storage and Potential Hydropower from Conduits." *Energy.gov*, February 2015. http://energy.gov/sites/prod/files/2015/06/f22/pumped-storage-potential-hydropower-from-conduits-final.pdf.

17. Christian-Smith, Juliet, and Laura Wisland. "Clean Energy Opportunities in California's Water Sector." Union of Concerned Scientists, April 2015. http://www.ucsusa.org/sites/default/files/attach/2015/04/clean-energy-opportunities-in-california-water-sector.pdf.

18. Clarke, Chris. "Explainer: Base Load and Peak Power." *KCET*, July 5, 2012. https://www.kcet.org/redefine/explainer-base-load-and-peaking-power.

19. U.S. Energy Information Administration (EIA). "Electric Power Annual 2009." 48. April 2011. http://www.eia.gov/electricity/annual/archive/03482009.pdf.

20. Open Energy Information (OEI). "Transparent Cost Database—Capacity Factor." Accessed December 29, 2016. http://en.openei.org/apps/TCDB.

21. International Renewable Energy Agency. "Hydropower." 18. June 2012. https://www.irena.org/DocumentDownloads/Publications/RE_Technologies_Cost_Analysis-HYDROPOWER.pdf.

22. U.S. Energy Information Administration. "Updated Capital Cost Estimates for Utility Scale Electricity Generating Plants." 6. April 2013. http://www.eia.gov/forecasts/capitalcost/pdf/updated_capcost.pdf.

23. Hadjerioua, Boualem, Yaxing Wei, and Shih-Chieh Kao. "An Assessment of Energy Potential at Non-Powered Dams in the United States." U.S. Department of Energy. viii. April 2012. http://nhaap.ornl.gov/sites/default/files/NHAAP_NPD_FY11_Final_Report.pdf.

24. Hadjerioua, Boualem, Yaxing Wei, and Shih-Chieh Kao. "An Assessment of Energy Potential at Non-Powered Dams in the United States." Oak Ridge National Laboratory, April 2012. http://nhaap.ornl.gov/sites/default/files/NHAAP_NPD_FY11_Final_Report.pdf.

25. Arizona Power Authority. "Hoover Dam." Accessed December 29, 2016. http://www.powerauthority.org/hoover-dam.

26. Therrien, Jean, and Gilles Bourgeois. "Fish Passage at Small Hydro Sites." The International Energy Agency. 13. March 2000. http://www.ieahydro.org/media/42697113/Fish%20Passage%20at%20Small-hydro%20Sites.pdf.

27. Ibid.; Collins, Gerald B. "Effects of Dams on Pacific Salmon and Steelhead Trout." 45. November 1976. http://spo.nmfs.noaa.gov/mfr3811/mfr38116.pdf.

28. Federal Energy Regulatory Commission (FERC). "Renewable Energy Production Tax Credit (PTC) Certification Orders Issued since 2005." Accessed December 29, 2016. https://www.ferc.gov/industries/hydropower/gen-info/comp-admin/tax-credit.pdf.

29. U.S. Department of Energy. "Pumped Storage and Potential Hydropower from Conduits." 23. February 2015. http://energy.gov/sites/prod/files/2015/06/f22/pumped-storage-potential-hydropower-from-conduits-final.pdf; U.S. Department of Energy. "Renewable Energy Production Tax Credit." Accessed December 29, 2016. http://energy.gov/savings/renewable-electricity-production-tax-credit-ptc.

30. United States. Congress. House. Hydropower Regulatory Efficiency Act of 2013. 2013. Accessed December 29, 2016. https://www.gpo.gov/fdsys/pkg/CRPT-113hrpt6/html/CRPT-113hrpt6.htm.

31. Kleinschmidt. "Maine Hydropower Study." 3–16. February 2015. http://www.maine.gov/energy/publications_information/001%20ME%20GEO%20Rpt%2002-04-15.pdf.

32. Oak Ridge National Laboratory, National Hydropower Association, and Hydro Research Foundation. "Small Hydropower Technology." 6. April 7, 2010. http://www.hydro.org/wp-content/uploads/2011/09/SmallHydroSummit.pdf.

33. U.S. Department of Energy. "Pumped Storage and Potential Hydropower from Conduits." 23. April 2015. http://energy.gov/sites/prod/files/2015/06/f22/pumped-storage-potential-hydropower-from-conduits-final.pdf.

34. Hydropower Reform Coalition. "Renewable Portfolio Standard (RPS)." Accessed December 29, 2016. http://www.hydroreform.org/policy/rps.

35. Stori, Val. "Environmental Rules for Hydropower in State Renewable Portfolio Standards." Clean Energy States Alliance, April 2013. http://www.cesa.org/assets/2013-Files/RPS/Environmental-Rules-for-Hydropower-in-State-RPS-April-2013-final-v2.pdf.

36. Haugen, Dan. "Renewable or Not? How States Count Hydropower." *Midwest Energy News*, January 13, 2012. http://midwestenergynews.com/2012/01/13/renewable-or-not-how-states-count-hydropowe.

37. Kubiszekski, Ida. "Federal Power Act of 1920, United States." *The Encyclopedia of Earth*, September 4, 2008. http://www.eoearth.org/view/article/152749.

38. United States. Congress. National Historic Preservation Act of 1966, As amended through 2006. 19. 2006. Accessed December 29, 2016. http://www.ncshpo.org/nhpa2008-final.pdf.

39. Hydropower Reform Coalition. "Laws Governing Hydropower Licensing." Accessed December 29, 2016. http://www.hydroreform.org/resources/laws.

40. Ibid.

41. U.S. Department of Energy. "Guidance for EPAct 2005 Section 242 Program." 1. Accessed December 29, 2016. http://energy.gov/sites/prod/files/2015/01/f19/Final%20Guidance%20for%20EPAct%202005%20Section%20242%20Hydroelectric%20Incentive%20Program.pdf.

42. Federal Register. "Guidance and Application for Hydroelectric Incentive Payments." April 26, 2016. https://www.federalregister.gov/documents/2016/04/26/2016-09700/guidance-and-application-for-hydroelectric-incentive-payments.

43. U.S. Department of Energy. "Final Guidance for EPAct 2005 Section 242 Hydroelectric Incentive Program." Accessed December 29, 2016. http://energy.gov

/eere/water/downloads/final-guidance-epact-2005-section-242-hydroelectric-incentive-program.

44. United States Congress. Energy Policy Act of 2005. 86. August 8, 2015. Accessed December 29, 2016. https://energy.gov/sites/prod/files/2014/03/f14/EPAof2005.pdf.

45. Thornton, Deborah. "Water, Water Everywhere but Not a Drop for Power." Public Interest Institute. 12. December 2012. http://limitedgovernment.org/publications/pubs/studies/ps-12-12.pdf.

46. Morrissey, Shannon. "FERC and USACE: The Necessity of Coordination in Implementation of the Hydropower Regulatory Efficiency Act." *UC Davis Law Review* 48, no. 4 (2015): 1581–1616. http://lawreview.law.ucdavis.edu/issues/48/4/Note/48-4_Morrissey.pdf.

47. Miles, Ann. "Hearing on Discussion Drafts Addressing Hydropower Regulatory Modernization and FERC Process Coordination under the Natural Gas Act." Federal Energy Regulatory Commission. 2–3. May 13, 2015. http://www.ferc.gov/CalendarFiles/20150513110741-Miles-testimony-05-13-2015.pdf.

48. Morrissey, Shannon. "FERC and USACE: The Necessity of Coordination in Implementation of the Hydropower Regulatory Efficiency Act." *UC Davis Law Review* 48, no. 4 (2015): 1581–1616. http://lawreview.law.ucdavis.edu/issues/48/4/Note/48-4_Morrissey.pdf.

49. Federal Energy Regulatory Commission Office of Energy Projects. "Hydropower Licensing—Get Involved: A Guide for the Public." 6. Accessed December 29, 2016. http://www.ferc.gov/resources/guides/hydro-guide.pdf.

50. Ibid., 6–7.

51. Ibid., 7.

52. Morrissey, Shannon. "FERC and USACE: The Necessity of Coordination in Implementation of the Hydropower Regulatory Efficiency Act." *UC Davis Law Review* 48, no. 4 (2015): 1581–1616. http://lawreview.law.ucdavis.edu/issues/48/4/Note/48-4_Morrissey.pdf.

53. Ibid.

54. Hydropower Regulatory Efficiency Act of 2013. 2. H.R. 267, 113th Cong., August 9, 2013. https://www.congress.gov/bill/113th-congress/house-bill/267.

55. U.S. Energy Information Administration. "Washington." Accessed December 29, 2016. http://www.eia.gov/state/?sid=WA.

56. Spiegel & McDiarmid LLP. "New Hydropower Legislation." September 11, 2013. http://www.spiegelmcd.com/files/Client%20Alert%20on%20New%20Hydropower%20Legislation_2013_09_11_03_40_20.pdf.

57. Federal Energy Regulatory Commission. "H-1 and Update on the Hydropower Regulatory Efficiency Act of 2013." 4. June 18, 2015. https://www.ferc.gov/industries/hydropower/indus-act/efficiency-act/H-1-presentation.pdf.

58. R. Bell. Conduit Exemptions at Federal Energy Regulatory Commission. Personal communication, October 19, 2015.

59. Spangler, Matthew. "E Hydropower Regulatory Efficiency Act of 2013: A Misguided Attempt to Bring Hydropower to the Forefront of the Renewable Energy

Landscape." *North Carolina Journal of Law & Technology*, October 8, 2013. http://ncjolt.org/the-hydropower-regulatory-efficiency-act-of-2013-a-misguided-attempt-to-bring-hydropower-to-the-forefront-of-the-renewable-energy-landscape.

60. Barrasso Senate Newsroom. "Barrasso-Risch Hydropower, Rural Jobs Bill Signed into Law." August 10, 2013. http://www.barrasso.senate.gov/public/index.cfm/news-releases?ID=68fce894-ae48-c8e4-9da6-e4fd546e309a.

61. U.S. Congress. House. Committee on Natural Resources. Bureau of Reclamation Small Conduit Hydropower Development and Rural Jobs Act (to accompany H.R. 678). House of Representatives. 5. March 25, 2013, H. Rep. http://www.gpo.gov/fdsys/pkg/CRPT-113hrpt24/pdf/CRPT-113hrpt24.pdf.

62. Ibid., 3.

63. Ibid., 2.

64. Barrasso Senate Newsroom. "Barrasso-Risch Hydropower, Rural Jobs Bill Signed into Law." August 10, 2013. http://www.barrasso.senate.gov/public/index.cfm/news-releases?ID=68fce894-ae48-c8e4-9da6-e4fd546e309a.

Chapter Nine

1. U.S. Nuclear Regulatory Commission. "The Nuclear Regulatory Commission Fact Sheet." March 29, 2012. http://www.nrc.gov/reading-rm/doc-collections/nuregs/brochures/br0099/r10/#functions.

2. Nuclear Energy Institute. "Nuclear Waste Fund Fee Suspended." http://www.nei.org/Master-Document-Folder/Multimedia/Infographics-Database/Nuclear-Waste-Fund-Fee-Suspended.

3. U.S. Environmental Protection Agency. "Public Health and Environmental Radiation Protection Standards for Yucca Mountain, Nevada (40 CFD Part 197): Rule History." https://www.epa.gov/radiation/public-health-and-environmental-radiation-protection-standards-yucca-mountain-nevada-40#rule-history.

4. Eureka County Nuclear Waste Office. "Timeline: 1954, 2014, Nuclear Waste Policy Dilemma." http://www.yuccamountain.org/time.htm.

5. Lander County Nuclear Waste Program. "History of Yucca Mountain Project." http://www.landercountynwop.com/historical.htm.

6. U.S. Department of Energy. Office of Civilian Radioactive Waste Management. "Final Supplemental Environmental Impact Statement for a Geologic Repository for the Disposal of Spent Nuclear Fuel and High-Level Radioactive Waste at Yucca Mountain, Nye County, Nevada." June 2008. S40–S42. http://energy.gov/sites/prod/files/EIS-0250-S1-FEIS-Summary-2008.pdf and U.S. Environmental Protection Agency. "Public Health and Environmental Radiation Protection Standards for Yucca Mountain, Nevada (40 CFD Part 197): Rule History." https://www.epa.gov/radiation/public-health-and-environmental-radiation-protection-standards-yucca-mountain-nevada-40#rule-history.

7. Office of U.S. Senator Harry Reid. "Yucca." http://www.reid.senate.gov/issues/yucca#.WCorLfkrLIU.

8. U.S. Department of Energy. Office of Civilian Radioactive Waste Management. "Final Supplemental Environmental Impact Statement for a Geologic Repository for the Disposal of Spent Nuclear Fuel and High-Level Radioactive Waste at Yucca Mountain, Nye County, Nevada." June 2008. S40–S42. http://energy.gov/sites/prod/files/EIS-0250-S1-FEIS-Summary-2008.pdf.

9. U.S. Environmental Protection Agency. "Public Health and Environmental Radiation Protection Standards for Yucca Mountain, Nevada (40 CFD Part 197): Rule History." https://www.epa.gov/radiation/public-health-and-environmental-radiation-protection-standards-yucca-mountain-nevada-40#rule-history.

10. Ibid.

11. U.S. Government Accountability Office. "Commercial Nuclear Waste: Effects of a Termination of the Yucca Mountain Repository Program and Lessons Learned." April 2011. http://www.gao.gov/assets/320/317627.pdf.

12. Ibid.

13. Ibid.

14. The Federal Government's Responsibilities and Liabilities under the Nuclear Waste Policy Act. Hearing before the Subcommittee on Environment and the Economy. U.S. House of Representatives. (2015) (testimony of Kim Cawley). https://www.cbo.gov/sites/default/files/114th-congress-2015-2016/reports/51035-NuclearWaste_Testimony.pdf.

15. U.S. Government Accountability Office. "Disposal of High-Level Nuclear Waste." http://www.gao.gov/key_issues/disposal_of_highlevel_nuclear_waste/issue_summary#t=0.

16. U.S. Government Accountability Office. "Commercial Nuclear Waste: Effects of a Termination of the Yucca Mountain Repository Program and Lessons Learned." April 2011. http://www.gao.gov/assets/320/317627.pdf.

17. Urban, Peter. "Reid Keeps New Yucca Mountain Funding at Bay." *Las Vegas Review-Journal*, December 18, 2015. http://www.reviewjournal.com/politics/government/reid-keeps-new-yucca-mountain-funding-bay.

18. U.S. Government Accountability Office. "Disposal of High-Level Nuclear Waste." http://www.gao.gov/key_issues/disposal_of_highlevel_nuclear_waste/issue_summary#t=0.

19. Nuclear Energy Institute. "Top 10 Facts about Yucca Mountain." http://www.nei.org/News-Media/News/News-Archives/Top-10-Facts-About-Yucca-Mountain.

20. U.S Energy Information Administration. "What Is U.S. Electricity Generation by Energy Source?" April 2016. https://www.eia.gov/tools/faqs/faq.cfm?id=427&t=3.

21. International Atomic Energy Agency. "Nuclear Share of Electricity Generation in 2015." April 30, 2016. https://www.iaea.org/PRIS/WorldStatistics/NuclearShareofElectricityGeneration.aspx.

22. Desai, Sachin, and Kathleen Schroeder. "U.S. Nuclear Foreign Ownership Policy Ready for Refreshed Interpretation." The Energy Bar Association,

May 16, 2016. http://www.felj.org/sites/default/files/docs/elj371/21-85-134-Desai_FINAL.pdf and U.S. Energy Information Administration. "How Much Electricity Does an American Home Use?" October 18, 2016. https://www.eia.gov/tools/faqs/faq.cfm?id=97&t=3.

23. Nuclear Energy Institute. "Quick Facts: Nuclear Energy in America." July 2016. http://www.nei.org/Master-Document-Folder/Backgrounders/Fact-Sheets/Quick-Facts-Nuclear-Energy-in-America.

24. Arizona Power Authority. "Hoover Dam." http://www.powerauthority.org/hoover-dam.

25. U.S. Energy Information Administration. "How Many Nuclear Power Plants Are in the United States, and Where Are They Located?" August 2, 2016. https://www.eia.gov/tools/faqs/faq.cfm?id=207&t=3.

26. Observations on the Key Attributes and Challenges of Storage and Disposal Options. Testimony before the Subcommittee on Energy and Water Development and Related Agencies. Hearing on "Commercial Spent Nuclear Fuel." U.S. House of Representatives. (2013) (testimony of Frank Rusco). http://www.gao.gov/assets/660/653731.pdf.

27. U.S. Nuclear Regulatory Commission. "Power Reactors." June 28, 2016. http://www.nrc.gov/reactors/power.html.

28. U.S. Nuclear Regulatory Commission. "Pressurized Water Reactors." January 15, 2015. http://www.nrc.gov/reactors/pwrs.html.

29. U.S. Nuclear Regulatory Commission. "Boiling Water Reactors." January 15, 2015. http://www.nrc.gov/reactors/bwrs.html.

30. U.S. Energy Information Administration. "How Many Nuclear Power Plants Are in the United States, and Where Are They Located?" August 2, 2016. https://www.eia.gov/tools/faqs/faq.cfm?id=207&t=3.

31. Beaver, William. "The Failed Promise of Nuclear Power." *The Independent Review* 15, no. 3 (2011): 399–411.

32. Ibid.

33. U.S. Nuclear Regulatory Commission. "Emergency Preparedness & Response: History." March 30, 2016. http://www.nrc.gov/about-nrc/emerg-preparedness/history.html.

34. World Nuclear Association. "Processing of Used Nuclear Fuel." November 2016. http://www.world-nuclear.org/information-library/nuclear-fuel-cycle/fuel-recycling/processing-of-used-nuclear-fuel.aspx.

35. Ibid.

36. Shughart, William F. "Why Doesn't U.S. Recycle Nuclear Fuel?" *Forbes*, October 1, 2014. http://www.forbes.com/sites/realspin/2014/10/01/why-doesnt-u-s-recycle-nuclear-fuel/#4947ff067db4.

37. Union of Concerned Scientists. "Nuclear Reprocessing: Dangerous, Dirty, and Expensive." http://www.ucsusa.org/nuclear-power/nuclear-plant-security/nuclear-reprocessing#.WCpIXfkrLIU.

38. Bunn, Matthew, Steve Fetter, John P. Holdren, and Bob van der Zwaan. "The Economics of Reprocessing vs. Direct Disposal of Spent Nuclear Fuel."

December 2003. Belfer Center for Science and International Affairs. http://belfercenter.ksg.harvard.edu/files/repro-report.pdf.

39. U.S. Nuclear Regulatory Commission. "Nuclear Regulatory Commission." October 17, 2016. http://www.nrc.gov/about-nrc/organization/nrrfuncdesc.html.

40. U.S. Nuclear Regulatory Commission. "Licensing." May 20, 2015. http://www.nrc.gov/about-nrc/regulatory/licensing.html.

41. U.S. Nuclear Regulatory Commission. "General Questions about NRC Fees." http://www.nrc.gov/about-nrc/regulatory/licensing/general-fee-questions.pdf.

42. U.S. Nuclear Regulatory Commission. Rule. "Revision of Fee Schedules; Fee Recovery for Fiscal Year 2015." *Federal Register* 80 (June, 30, 2015): 37431–64. https://www.federalregister.gov/documents/2015/06/30/2015-15763/revision-of-fee-schedules-fee-recovery-for-fiscal-year-2015.

43. U.S. Nuclear Regulatory Commission. "Nuclear Power Plant Licensing Process." http://www.nrc.gov/reading-rm/doc-collections/nuregs/brochures/br0298/br0298r2.pdf.

44. Ibid.

45. Ibid.

46. Ibid.

47. U.S. Nuclear Regulatory Commission. "Frequently Asked Questions about License Applications for New Nuclear Power Reactors." December 2009. http://www.nrc.gov/reading-rm/doc-collections/nuregs/brochures/br0468/br0468.pdf.

48. Ross, Jerry, and Barry M. Staw. "Organizational Escalation and Exit: Lessons from the Shoreham Nuclear Power." *The Academy of Management Journal* 36, no. 4 (1993): 701–32.

49. The Federal Government's Responsibilities and Liabilities under the Nuclear Waste Policy Act. Hearing before the Subcommittee on Environment and the Economy. U.S. House of Representatives. (2015) (testimony of Kim Cawley). https://www.cbo.gov/sites/default/files/114th-congress-2015-2016/reports/51035-NuclearWaste_Testimony.pdf.

50. Ibid.

51. Ibid.

52. "Court Orders Halt to Nuclear Waste Fees." *World Nuclear News*, November 20, 2013. http://www.world-nuclear-news.org/WR-Court-orders-halt-to-nuclear-waste-fees-2011134.html.

53. U.S. Government Accountability Office. "Commercial Nuclear Waste: Effects of a Termination of the Yucca Mountain Repository Program and Lessons Learned." April 2011. http://www.gao.gov/assets/320/317627.pdf.

54. Ibid.

Chapter Ten

1. Arizona Rural Policy Institute. "Demographic Analysis of the Navajo Nation Using 2010 Census and 2010 American Community Survey Estimates." 2010. http://azcia.gov/Documents/Links/DemoProfiles/Navajo%20Nation.pdf.

2. Salt River Project. "Navajo Generating Station: Lease Status." https://www.ngspower.com/about/lease.aspx.

3. Hammond, Joseph. "Navajo Nation in Crisis as EPA Tries to Shutter the West's Largest Coal Plan." *The Daily Caller*, May 3, 2016. http://dailycaller.com/2016/05/03/navajo-nation-in-crisis-as-epa-tries-to-shutter-the-wests-largest-coal-plant.

4. Ibid.

5. Salt River Project. "Navajo Generating Station: Lease Status." https://www.ngspower.com/about/lease.aspx and Peabody Energy. "Fact Sheet: Kayenta Mine." August 22, 2016. https://mscusppegrs01.blob.core.windows.net/mmfiles/files/factsheets/kayenta.pdf.

6. The Shift Project. "Breakdown of Electricity Generation by Source." http://www.tsp-data-portal.org/Breakdown-of-Electricity-Generation-by-Energy-Source#tspQvChart.

7. U.S. Energy Information Administration. "Direct Federal Financial Interventions and Subsidies in Energy in Fiscal Year 2013." March 2015. xix–xxi. http://www.eia.gov/analysis/requests/subsidy/pdf/subsidy.pdf.

8. U.S. Energy Information Administration. "What Is U.S. Electricity Generation by Energy Source?" April 1, 2016. https://www.eia.gov/tools/faqs/faq.cfm?id=427&t=3.

9. U.S. Energy Information Administration. "Coal Explained: How Much Coal Is Left." June 17, 2016. http://www.eia.gov/energyexplained/index.cfm?page=coal_reserves.

10. American Electric Power. "Pulverized Coal Technologies." http://www.aep.com/about/IssuesAndPositions/Generation/Technologies/PulverizedCoal.aspx.

11. Rubin, Edward S., Michael B. Berkenpas, H. Christopher Frey, Chao Chen, Sean T. McCoy, and Constance J. Zaremsky. "Technical Documentation: Integrated Gasification Combined Cycle Systems (IGCC) with Carbon Capture and Storage (CCS)." 3–4. http://repository.cmu.edu/cgi/viewcontent.cgi?article=1073&context=ep.

12. U.S. Energy Information Administration. "Coal Explained: How Much Coal Is Left." June 17, 2016. http://www.eia.gov/energyexplained/index.cfm?page=coal_reserves.

13. U.S. Department of Energy. "The Clean Coal Technology Program." February 12, 2013. http://www.fossil.energy.gov/education/energylessons/coal/coal_cct2.html and U.S. Environmental Protection Agency. "Carbon Dioxide Capture and Sequestration." http://www3.epa.gov/climatechange/ccs.

14. Martinson, Erica. "The Fall of Coal." *Politico*, April 16, 2015. http://www.politico.com/story/2015/04/coal-power-plants-epa-regulations-117011.html.

15. Restuccia, Andrew. "Michael Bloomberg's War on Coal." *Politico*, April 8, 2015. http://www.politico.com/story/2015/04/michael-bloomberg-environment-coal-sierra-club-116793.html.

16. Sierra Club. "About Us." Accessed June 8, 2015. http://content.sierraclub.org/coal/about-the-campaign.

17. Ibid.

18. U.S. Energy Information Administration. "Direct Federal Financial Interventions and Subsidies in Energy in Fiscal Year 2013." March 2015. xix–xxi. http://www.eia.gov/analysis/requests/subsidy/pdf/subsidy.pdf.

19. Litterman, Bob. "What Is the Right Price for Carbon Emissions?" *Regulation* 36, no. 2 (2013): 38–43.

20. Pindyck, Robert. S. "Climate Change Policy: What Do the Models Tell Us?" July 2013. National Bureau of Economic Research Working Paper 19244. http://web.mit.edu/rpindyck/www/Papers/Climate-Change-Policy-What-Do-the-Models-Tell-Us.pdf.

21. U.S. Energy Information Administration. "Power Plant Emissions of Sulfur Dioxide and Nitrogen Oxides Continue to Decline in 2012." February 27, 2013. http://www.eia.gov/todayinenergy/detail.cfm?id=10151.

22. U.S. Environmental Protection Agency. "What Are the Biggest Sources of Mercury Air Emissions in the U.S.?" https://publicaccess.zendesk.com/hc/en-us/articles/211395308-What-are-the-biggest-sources-of-mercury-air-emissions-in-the-U-S-.

23. U.S. Environmental Protection Agency. "Mercury: Health Effects." 2015. http://www.epa.gov/mercury/effects.htm.

24. U.S. Environmental Protection Agency. "What Is the EPA Doing to Reduce Emissions of Mercury into the Air?" 2015. http://publicaccess.supportportal.com/link/portal/23002/23012/Article/21820/What-is-EPA-doing-to-reduce-emissions-of-mercury-into-the-air.

25. U.S. Environmental Protection Agency. "Healthier Americans." 2015. http://www3.epa.gov/mats/health.html.

26. Sabas, Matthew. "EPA Rules Bring Pink Slips to a Town Near You." April 28, 2015. Institute for Energy Research. http://instituteforenergyresearch.org/analysis/epa-continues-all-pain-no-gain-policies-with-mercury-and-air-toxic-standards.

27. Michigan et al. v. Environmental Protection Agency et al. 576 U.S. ____ (2015).

28. Wolff, Eric. "Supreme Court's Eventual MATS Ruling Will Be (Mostly) Moot. *SNL Data Dispatch*, May 14, 2015. https://www.snl.com/InteractiveX/Article.aspx?cdid=A-32620730-13109.

29. Martinson, Erica. "The Fall of Coal." *Politico*, April 16, 2015. http://www.politico.com/story/2015/04/coal-power-plants-epa-regulations-117011.html.

30. Ibid.

31. Taxpayers for Common Sense. "Coal: A Long History of Subsidies." 2009. http://www.taxpayer.net/images/uploads/downloads/Coal_subsidies_factsheet_2009.pdf.

32. U.S. Energy Information Administration. "Direct Federal Financial Interventions and Subsidies in Energy in Fiscal Year 2013." March 2015. 11. http://www.eia.gov/analysis/requests/subsidy/pdf/subsidy.pdf.

33. Tax Policy Center. "What Is the Tax Expenditure Budge?" http://www.taxpolicycenter.org/briefing-book/background/shelters/expenditures.cfm.

34. Ibid.

35. Taxpayers for Common Sense. "Coal: A Long History of Subsidies." 2009. http://www.taxpayer.net/images/uploads/downloads/Coal_subsidies_factsheet_2009.pdf.

36. U.S. Internal Revenue Service. "$1 Billion in Tax Credits Allocated to Clean Coal Projects." November 30, 2006. https://www.irs.gov/uac/1-billion-in-tax-credits-allocated-to-clean-coal-projects.

37. Energy Improvement and Extension Act of 2008. Public Law 110-343. *U.S. Statutes at Large* 122 (2008): 3807–61.

38. Pfund, Nancy, and Ben Healey. "What Would Jefferson Do? The Historical Role of Federal Subsidies in Shaping America's Energy Future." September 2011. DBL Partners. 9. http://www.dblinvestors.com/documents/What-Would-Jefferson-Do-Final-Version.pdf.

39. U.S. Office of Inspector General. "Coal Management Program, U.S. Department of the Interior." June 2013. https://www.doioig.gov/sites/doioig.gov/files/CR-EV-BLM-0001-2012Public.pdf.

40. U.S. Government Accountability Office. "Coal Leasing: BLM Could Enhance Appraisal Process, More Explicitly Consider Coal Exports, and Provide More Public Information." December 2013. GAO-14-140. http://www.gao.gov/assets/660/659801.pdf

41. U.S. Office of Inspector General. "Coal Management Program, U.S. Department of the Interior." June 2013. https://www.doioig.gov/sites/doioig.gov/files/CR-EV-BLM-0001-2012Public.pdf.

42. Taxpayers for Common Sense. "Coal: A Long History of Subsidies." 2009. http://www.taxpayer.net/images/uploads/downloads/Coal_subsidies_factsheet_2009.pdf.

43. Office of Fossil Energy. "Clean Coal Power Initiative." http://energy.gov/fe/science-innovation/clean-coal-research/major-demonstrations/clean-coal-power-initiative.

44. Taxpayers for Common Sense. "Coal: A Long History of Subsidies." 2009. http://www.taxpayer.net/images/uploads/downloads/Coal_subsidies_factsheet_2009.pdf.

45. Office of Fossil Energy. "Clean Coal Power Initiative." http://energy.gov/fe/science-innovation/clean-coal-research/major-demonstrations/clean-coal-power-initiative.

46. Institute for Energy Research. "Energy Regulation in the States: A Wake-up Call." Appendices A and B. http://instituteforenergyresearch.org/states.

47. Ibid.

48. Ibid.

49. Patrick, Aron, Adam Blandford, and Leonard K. Peters. "The Vulnerability of the United States Economy to Electricity Price Increases." March 2015. Kentucky Energy and Environment Cabinet. 11, 18. http://energy.ky.gov/Programs/Data%20Analysis%20%20Electricity%20Model/Vulnerability_to_Electricity_Prices.pdf.

50. Institute for Energy Research. "Energy Regulation in the States: A Wake-up Call: Wyoming." http://instituteforenergyresearch.org/states and

Godby, Robert, Roger Coupal, David Taylor, and Tim Considine. "The Impact of the Coal Economy on Wyoming." February 2015. Center for Energy Economics and Public Policy. 4. http://www.uwyo.edu/cee/_files/docs/wia_coal_full-report.pdf.

51. Wyoming State Geologic Survey. "Wyoming Coal." 2015. http://www.wsgs.wyo.gov/energy/coal.

52. U.S. Environmental Protection Agency. "Clean Power Plan: State at a Glance: Wyoming." August 3, 2015. http://www3.epa.gov/airquality/cpptoolbox/wyoming.pdf.

53. Godby, Robert, Roger Coupal, David Taylor, and Tim Considine. "The Impact of the Coal Economy on Wyoming." February 2015. Center for Energy Economics and Public Policy. 93. http://www.uwyo.edu/cee/_files/docs/wia_coal_full-report.pdf.

54. Volcovici, Valerie. "Fifteen U.S. States Seek to Block EPA Carbon Rule." *Reuters*, August 13, 2015. http://www.reuters.com/article/2015/08/13/us-climate change-cleanpower-plan-idUSKCN0QI2D920150813 and Schwartz, Thea. "Vermont Joins 16 States and Cities in Announcing Their Intention to Oppose Request for Court Stay of the Clean Power Rule." August 14, 2015. Vermont Office of the Attorney General. http://ago.vermont.gov/focus/news/vermont-joins-16-states-and-cities-in-announcing-their-intention-to-oppose-request-for-court-stay-of-the-clean-power-rule.php.

55. Institute for Energy Research. "The Status of Renewable Electricity Mandates in the States." 1. http://instituteforenergyresearch.org/wp-content/uploads/2011/01/IER-RPS-Study-Final.pdf.

56. Long, Noah. "California Gets Closer to Closing the Door on Coal." *Natural Resources Defense Council Switchboard*, October 18, 2010. http://switchboard.nrdc.org/blogs/nlong/california_gets_closer_to_clos.html and California Energy Commission. "SB 1368 Emission Performance Standards." http://www.energy.ca.gov/emission_standards.

Chapter Eleven

1. New York Department of Environmental Conservation. "Final Supplemental Generic Environmental Impact Statement on the Oil, Gas, and Solution Mining Regulatory Program: Regulatory Program for Horizontal Drilling and High-Volume Hydraulic Fracturing to Develop the Marcellus Shale and Other Low-Permeability Gas Reservoirs Findings Statement." June 2015. http://www.dec.ny.gov/docs/materials_minerals_pdf/findingstatehvhf62015.pdf.

2. New York Department of Environmental Conservation. "Final Supplemental Generic Environmental Impact Statement, Executive Summary." 2. http://www.dec.ny.gov/docs/materials_minerals_pdf/fsgeis2015es.pdf.

3. U.S. Energy Information Administration. "New York Electricity Profile 2013." https://www.eia.gov/electricity/state/newyork and Gerken, James. "Gov. Andrew Cuomo to Ban Fracking in New York State." *The Huffington Post*, December 17, 2014. http://www.huffingtonpost.com/2014/12/17/cuomo-fracking-new-york-state_n_6341292.html.

4. New York Department of Environmental Conservation. "New York State Officially Prohibits High-Volume Hydraulic Fracturing: DEC Issues Findings Statement Concluding Extensive Seven-Year Review." June 29, 2015. http://www.dec.ny.gov/press/102337.html.

5. U.S. Environmental Protection Agency. "Assessment of the Potential Impacts of Hydraulic Fracturing for Oil and Gas on Drinking Water Resources." June 2015. EPA/600/R-15/047a. https://www.epa.gov/sites/production/files/2015-07/documents/hf_es_erd_jun2015.pdf.

6. McCarthy, Gina. "U.S.-China Clean Air and Climate Cooperation." *C-SPAN Video,* December 2, 2013. https://www.c-span.org/video/?316536-1/top-epa-chief-uschina-climate-change-gains.

7. Obama, Barack. "Remarks by President Obama in Press Conference after GCC Summit" (speech, Camp David, May 14, 2015). https://www.whitehouse.gov/the-press-office/2015/05/14/remarks-president-obama-press-conference-after-gcc-summit.

8. Miller, Craig. "Interior Secretary: Local Fracking Bans Are 'Wrong Way To Go.'" *KQED Science*, January 2, 2015. https://ww2.kqed.org/science/2015/01/02/interior-secretary-local-fracking-bans-are-wrong-way-to-go.

9. Harder, Amy, and Erica Orden. "Bloomberg Criticizes New York Fracking Ban." *The Wall Street Journal*, April 8, 2015. http://www.wsj.com/articles/bloomberg-criticizes-new-york-fracking-ban-1428520873.

10. Brooks, Jon. "Jerry Brown Lashes Out against Environmentalists over Fracking." *KQED News*, October 30, 2013. https://ww2.kqed.org/news/2013/10/30/jerry-brown-fracking.

11. U.S. Energy Information Administration. "New York State Profile and Energy Estimates." http://www.eia.gov/state/?sid=NY#tabs-4.

12. U.S. Energy Information Administration. "New York State Profile and Energy Estimates." http://www.eia.gov/state/?sid=NY#tabs-4.

13. Waldman, Scott, and Bill Mahoney. "New York Increasingly Reliant on Natural Gas for Heat." *Politico*, October 5, 2015. http://www.politico.com/states/new-york/albany/story/2015/10/new-york-increasingly-reliant-on-natural-gas-for-heat-026345.

14. Yerman, Marcia G. "Fracking New York: It's Not Over 'Till It's Over." *The Huffington Post*, May 1, 2015. http://www.huffingtonpost.com/marcia-g-yerman/fracking-new-york-its-not_b_7183342.html.

15. Mathias, Christopher. "New York Towns Threaten Secession over Cuomo's Ban on Fracking." *The Huffington Post*, February 2, 2015. http://www.huffingtonpost.com/2015/02/20/new-york-fracking-secession-southern-tier-cuomo_n_6722296.html.

16. Conca, James. "New York Fracking Ban Contrary to State's Energy Future." *Forbes,* December 27, 2016. http://www.forbes.com/sites/jamesconca/2014/12/27/new-york-fracking-ban-contrary-to-states-energy-future.

17. U.S. Energy Information Administration. "What Is U.S. Electricity Generation by Energy Source?" April 1, 2016. https://www.eia.gov/tools/faqs/faq.cfm?id=427&t=3.

18. U.S. Energy Information Administration. "Direct Federal Financial Interventions and Subsidies in Energy in Fiscal Year 2013." March 2015. xix–xxi. http://www.eia.gov/analysis/requests/subsidy/pdf/subsidy.pdf.

19. MacRae, Michael. "Fracking: A Look Back." December 2012. The American Society of Mechanical Engineers. https://www.asme.org/engineering-topics/articles/fossil-power/fracking-a-look-back.

20. U.S. Energy Information Administration. "June 2015 Monthly Energy Review." June 2015. 105. http://www.eia.gov/totalenergy/data/monthly/archive/00351506.pdf.

21. American Oil and Gas Historical Society. "Shooters—A 'Fracking' History." http://aoghs.org/technology/hydraulic-fracturing.

22. American Exploration and Production Council. "The Real Facts about Fracture Stimulation: The Technology behind America's New Natural Gas Supplies." February 10, 2010. http://energyindepth.org/wp-content/uploads/2011/10/Real-facts-behind-fracture-stimulation-technology.pdf.

23. Wang, Zhongmin, and Alan Krupnick. "US Shale Gas Development: What Led to the Boom?" May 2013. Resources for the Future Issue Brief 13-04. http://www.rff.org/files/sharepoint/WorkImages/Download/RFF-IB-13-04.pdf.

24. Hausman, Catherine, and Ryan Kellogg. "Welfare and Distributional Implications of Shale Gas." March 2015. Brookings Papers on Economic Activity. 1–3, 18. http://www.brookings.edu/~/media/projects/bpea/spring-2015/2015a_hausman.pdf.

25. Steitz, Christoph, and Ernest Scheyder. "Special Report: How Fracking Helps America Beat German Industry." *Reuters*, June 2, 2014. http://www.reuters.com/article/2014/06/02/us-usa-germany-power-specialreport-idUSKBN0ED0CS20140602.

26. "The History of Regulation." September 20, 2013. Natural Gas Supply Association (NGSA). http://naturalgas.org/regulation/history.

27. U.S. Energy Information Administration. "Direct Federal Financial Interventions and Subsidies in Energy in Fiscal Year 2013." March 2015. xv–xvi. http://www.eia.gov/analysis/requests/subsidy/pdf/subsidy.pdf

28. Management Information Services, Inc. "60 Years of Energy Incentives; Analysis of Federal Expenditures for Energy Development." October 2011. 1. http://www.misi-net.com/publications/NEI-1011.pdf.

29. Ibid.

30. Federal Financial Support for Fuels and Energy Technologies. Hearing before the Subcommittee on Energy. U.S. House of Representatives. (2013) (testimony of Terry M. Dinan). https://www.cbo.gov/sites/default/files/113th-congress-2013-2014/reports/43993_EnergyTechnologies_one-column.pdf.

31. Management Information Services, Inc. "60 Years of Energy Incentives: Analysis of Federal Expenditures for Energy Development." October 2011. 1. http://www.misi-net.com/publications/NEI-1011.pdf.

32. U.S. Energy Information Administration. "Direct Federal Financial Interventions and Subsidies in Energy in Fiscal Year 2013." March 2015. xv–xvi. http://www.eia.gov/analysis/requests/subsidy/pdf/subsidy.pdf.

33. Ibid., 20.

34. Ibid., ix, 28; Catalog of Federal Domestic Assistance. “State Clean Diesel Grant Program.” Accessed October 31, 2015. https://www.cfda.gov/index?s=program&mode=form&tab=step1&id=482545356eaf971f401b6244ff7166c4.

35. U.S. Energy Information Administration. “Direct Federal Financial Interventions and Subsidies in Energy in Fiscal Year 2013.” March 2015. 43. http://www.eia.gov/analysis/requests/subsidy/pdf/subsidy.pdf.

36. Ibid., x, xv–xvi, 51.

37. U.S. Federal Energy Regulatory Commission. “What FERC Does.” June 24, 2014. http://www.ferc.gov/about/ferc-does.asp.

38. Wethe, David. “Drillers Take Second Crack at Fracking Wells to Cut Cost.” *Bloomberg*, February 9, 2015. http://www.bloomberg.com/news/articles/2015-02-10/drillers-take-second-crack-at-fracking-wells-to-cut-cost-energy.

39. Parfomak, Paul W. “Interstate Natural Gas Pipelines: Process and Timing of FERC Permit Application Review.” January 16, 2015. Congressional Research Service Report No. R43138. https://www.fas.org/sgp/crs/misc/R43138.pdf.

40. Center for Western Priorities. “A Fair Share: The Case for Updating Oil and Gas Royalties on Our Public Lands.” June 18, 2015. 2. http://www.westernpriorities.org/wp-content/uploads/2015/06/Royalties-Report_update.pdf.

41. U.S. Department of the Interior. “Interior Department Seeks Public Dialogue on Reform of Federal Onshore Oil and Gas Regulations.” April 17, 2015. Bureau of Land Management News Release. http://www.blm.gov/wo/st/en/info/newsroom/2015/april/interior_department.html.

42. U.S. Bureau of Land Management. Advance Notice of Proposed Rulemaking. “Oil and Gas Leasing; Royalty on Production, Rental Payments, Minimum Acceptable Bids, Bonding Requirements, and Civil Penalty Assessments.” *Federal Register* 80, no. 76 (April 21, 2015): 22148–56. https://www.gpo.gov/fdsys/pkg/FR-2015-04-21/pdf/2015-09033.pdf.

43. U.S. Bureau of Land Management. Proposed Rule. “Waste Prevention, Production Subject to Royalties, and Resource Conservation.” *Federal Register* 81, no. 25 (February 8, 2016): 6616–86. http://www.blm.gov/style/medialib/blm/wo/Communications_Directorate/public_affairs/news_release_attachments.Par.15043.File.dat/VF%20Proposed%20Rule%20Waste%20Prevention.pdf.

44. Ohio Environmental Protection Agency. “Understanding the Basics of Gas Flaring.” November 2014. http://www.epa.state.oh.us/Portals/27/oil%20and%20gas/Basics%20of%20Gas%20Flaring.pdf.

45. U.S. Environmental Protection Agency. “Overview of Greenhouse Gases.” https://www.epa.gov/ghgemissions/overview-greenhouse-gases.

46. Phillips, Susan. “Burning Question: What Would Life Be Like without the Halliburton Loophole?” *StateImpact,* December 5, 2011. http://stateimpact.npr.org/pennsylvania/2011/12/05/burning-question-what-would-life-be-like-without-the-halliburton-loophole.

47. Brady, William J. “Hydraulic Fracturing Regulation in the United States: The Laissez-Faire Approach of the Federal Government and Varying State Regulations.” 2012. University of Denver, Sturm College of Law and Grimshaw &

Harring, P.C. 3, 5. http://www.law.du.edu/documents/faculty-highlights/Intersol-2012-HydroFracking.pdf.

48. U.S. Energy Information Administration. "Rankings: Natural Gas Marketed Production." 2014. http://www.eia.gov/state/rankings/?sid=US#/series/47&CFID=20610564&CFTOKEN=243d553f77de02b6-F28D2A0B-25B3-1C83-54CE29AB30FA9864&jsessionid=8430504ace19cb649f88516672267e2e35b1 and U.S. Energy Information Administration. "Frequently Asked Questions: Which States Consume and Produce the Most Natural Gas?" November 9, 2015. http://www.eia.gov/tools/faqs/faq.cfm?id=46&t=8.

49. U.S. Energy Information Administration. "Review of Emerging Resources: U.S. Shale Gas and Shale Oil Plays." June 2011. 4. http://www.eia.gov/analysis/studies/usshalegas/pdf/usshaleplays.pdf.

50. Act 152 VT H.464. An act relating to hydraulic fracturing wells for natural gas and oil production. 2012. http://legislature.vermont.gov/bill/status/2012/H.464 and NewsCore. "Vermont Becomes First State to Ban Fracking." *Fox News*, May 17, 2012. http://www.foxnews.com/politics/2012/05/17/vermont-becomes-first-state-to-ban-fracking and KSE Focus. "States Take Wait and See Approach on Fracking Regulation." *CQ Roll Call*, July 9, 2015. http://congress.org/2015/07/09/states-take-wait-and-see-approach-on-fracking-regulation.

51. U.S. Energy Information Administration. "Natural Gas Gross Withdrawals and Production." October 30, 2015. http://www.eia.gov/dnav/ng/ng_prod_sum_a_EPG0_FGG_mmcf_a.htm and U.S. Energy Information Administration. "Vermont." May 21, 2015. https://www.eia.gov/state/analysis.cfm?sid=VT.

52. Environment-Hydraulic Fracturing-Regulations, MD SB0409. 2015. http://mgaleg.maryland.gov/webmga/frmMain.aspx?id=sb0409&stab=01&pid=billpage&tab=subject3&ys=2015RS; Environment-Hydraulic Fracturing-Regulations MD HB0449. 2015. http://mgaleg.maryland.gov/webmga/frmMain.aspx?id=HB0449&stab=01&pid=billpage&tab=subject3&ys=2015RS.

53. Wheeler, Timothy B. "O'Malley Administration Sets Out Path to Fracking in Md." *Baltimore Sun*, November 25, 2014. http://www.baltimoresun.com/features/green/blog/bs-md-fracking-regulation-20141125-story.html.

54. Express Preemption of Regulation of Oil and Gas, TX HB0040. 2015. http://www.legis.state.tx.us/BillLookup/History.aspx?LegSess=84R&Bill=HB0040 and Texas House Research Organization. "HRO Bill Analysis: HB 40." April 17, 2015. 4. http://www.hro.house.state.tx.us/pdf/ba84r/hb0040.pdf#navpanes=0.

55. Miller, Craig. "Interior Secretary: Local Fracking Bans Are 'Wrong Way To Go.'" *KQED Science*, January 2, 2015. http://ww2.kqed.org/science/2015/01/02/interior-secretary-local-fracking-bans-are-wrong-way-to-go.

56. Davis, Charles. "The Politics of 'Fracking': Regulating Natural Gas Drilling Practices in Colorado and Texas." *Review of Policy Research* 29, no. 2 (2012): 177–91.

57. Oil and gas; authorizing regulation by municipalities, counties and other political subdivisions; prohibiting certain regulations, OK SB809. 2015.

http://www.oklegislature.gov/BillInfo.aspx?Bill=SB809&Session=1500 and Murphy, Sean. "Oklahoma Joins Texas with Law Preventing Cities, Towns from Banning Oil and Gas Operations." *U.S. News & World Report,* May 29, 2015. http://www.usnews.com/news/business/articles/2015/05/29/oklahoma-is-latest-state-to-prevent-local-fracking-bans.

58. Ibid.

59. Penn State Public Broadcasting. "Explore Shale." August 2015. Pennsylvania State University & Colcom Foundation. http://www.exploreshale.org and Energy from Shale. "A Few Days of Fracking, Decades of Oil and Gas Production." 2015. http://www.energyfromshale.org/articles/few-days-fracking-decades-oil-and-gas-production.

60. Gold, Russell. *The Boom: How Fracking Ignited the American Energy Revolution and Changed the World* (New York: Simon & Schuster, 2015), 53–54.

61. Wertz, Joe. "State Seismologist: Oklahoma Earthquakes 'Very Likely' Triggered by Oil and Gas Disposal Wells." *StateImpact*, April 21, 2015. https://stateimpact.npr.org/oklahoma/2015/04/21/state-seismologist-oklahoma-earthquakes-very-likely-triggered-by-oil-and-gas-disposal-wells.

62. Ibid.; United States Geological Survey. "Earthquakes in the Stable Continental Region." January 2014. http://earthquake.usgs.gov/earthquakes/eventpage/usc000tca7#general_summary.

63. Bunzey, Rachael. "Natural Gas and Green Completion in a Nut Shell." *Energy in Depth*, November 26, 2012. http://energyindepth.org/marcellus/natural-gas-and-green-completion-in-a-nut-shell and IPIECA. "Green Completions." 2013. http://www.ipieca.org/energyefficiency/solutions/78161.

64. Soraghan, Mike. "Significant Quake Damages Buildings Near Okla. Pipeline Hub." *Energywire*, November 7, 2016. http://www.eenews.net/energywire/2016/11/07/stories/1060045348.

65. Fischetti, Mark. "Groundwater Contamination May End the Gas-Fracking Boom." *Scientific American*, September 1, 2013. http://www.scientificamerican.com/article/groundwater-contamination-may-end-the-gas-fracking-boom.

66. U.S. Environmental Protection Agency. "Assessment of the Potential Impacts of Hydraulic Fracturing for Oil and Gas on Drinking Water Resources (External Review Draft)." June 4, 2015. http://cfpub.epa.gov/ncea/hfstudy/recordisplay.cfm?deid=244651.

Index

About the Authors

Ryan M. Yonk, PhD, Georgia State University, is an assistant research professor in the Department of Economics and Finance at Utah State University and one of the founders of Strata Policy. Dr. Yonk is the author of numerous academic journal articles, policy reports, and books focused on the intersection of public policy, economics, and politics. Dr. Yonk writes regularly for *The Hill* and a number of other popular press outlets. His dissertation was on the political effects of quality of life.

Jordan Lofthouse, MSE, Utah State University, has published several op-eds in *The Hill* and *The Salt Lake Tribune*. He has worked on several policy reports relating to energy, the environment, and public lands. His current research areas include renewable energy, national park management, and federal environmental laws.

Megan Hansen, MSE, Utah State University, has published op-eds in *USA Today*, *Newsweek*, and *The Salt Lake Tribune* and has worked on numerous policy reports on energy and environmental issues as a policy analyst at Strata. Her research areas include electricity markets, small-hydropower regulation, and the Endangered Species Act.

www.ingramcontent.com/pod-product-compliance
Ingram Content Group UK Ltd.
Pitfield, Milton Keynes, MK11 3LW, UK
UKHW020843290926
15470UKWH00004B/1528

9 781440 853913